Debunked!

Debunked!

ESP, Telekinesis, and Other Pseudoscience

Georges Charpak and Henri Broch
translated by Bart K. Holland

The Johns Hopkins University Press
Baltimore and London

Although every effort has been made to ensure that the information in this book is correct, the publisher can give no assurance to that effect. Neither the authors nor the publisher accepts responsibility for loss, damage, injury, and/or death resulting from any act(s) of omission or commission in demonstrating any of the techniques described in the book

Originally published as *Devenez sorciers, devenez savants* © Éditions Odile Jacob, April 2002
Printed in the United States of America on acid-free paper
9 8 7 6 5 4 3 2 1

"Weak Sources of Natural Radiation" in chapter 3 is taken from Georges Charpak and Richard L. Garwin, *Megawatts and Megatons: A Turning Point in the Nuclear Age?* (New York: Knopf, 2001).

The Johns Hopkins University Press
2715 North Charles Street
Baltimore, Maryland 21218-4363
www.press.jhu.edu

Library of Congress Cataloging-in-Publication Data

Charpak, Georges.
    [Devenez sorciers, devenez savants English]
    Debunked! ESP, telekinesis, and other pseudoscience / Georges Charpak and Henri Broch ; translated by Bart K. Holland.
        p.   cm.
    ISBN 0-8018-7867-5 (hardcover : alk. paper)
1. Occultism and science.   I. Broch, Henri.   II. Title.
BF1409.5.C4313 2004
130—dc22   2003015032

A catalog record for this book is available from the British Library.

Contents

Translator's Preface

In my usual role as a medical school faculty member, my work centers on probability as an aid to clear thinking in epidemiology, clinical trials, and other areas. My two previous books from the Johns Hopkins University Press—*Probability without Equations: Concepts for Clinicians* and *What Are the Chances? Voodoo Deaths, Office Gossip, and Other Adventures in Probability*—grew out of this professional interest. Another of my interests, however, focuses on language and translation, which until recently seemed a quite disparate area. Then Trevor Lipscombe, editor in chief at Hopkins, suggested that I translate this book. I quickly assented, pleased to have an opportunity to unite my two interests in one work and to pursue a close reading of a book I'd want to read anyway. A book using probability and statistics to distinguish scientific findings from nonsense, especially one written by a Nobelist and his distinguished colleague, was surely not something I'd seen recently.

Any translation requires a careful balance between exactitude and readability. I haven't strayed too far from the French text, but I have made some changes to make this version readable. Most examples pertaining to places, institutions, incidents, and people in France have been retained because they are important examples and were the ones chosen by the authors; however, some adaptation was required. Occasionally, the significance of a place or a celebrity would probably not be recognized by those not living in France. In such cases I sometimes inserted a sentence or two to clarify the context. I dropped an example entirely when it would have required an extensive explanation interfering with the flow of the text and when the example was clearly only secondary support to the main points. Sometimes the rarity of an event was illustrated by the number of cases that one might expect to find in a country the size of France; I changed these examples to indicate the number found in the United States and made other, similar changes, as in a chart involving common first names. It is, however, appropriate to retain the flavor and the overwhelming majority of the content of the original work as presented by Charpak and Broch. Moreover, those interested in the rise of pseudoscience in the United States will want to see how psychokinesis, astrology, and the like are manifest in another setting. And that is why most examples have been retained.

We've made two other changes of a stylistic nature: First, footnotes containing explanations and references have been incorporated into the text itself. Second, very long sentences in a distinctive, glorious Gallic rhetorical style have been reduced in frequency by rewriting (although not eliminated completely). There comes a point where multiple subordinate clauses are not considered good writing (or good reading) in contemporary English, while their complexity may be considered admirable and lovely in French. I opted for intelligibility rather than the unalloyed preservation of style, striving for a version that would be considered as well written in my language as Charpak and Broch's is in theirs.

My linguistically oriented family gets the lion's share of credit for my knowledge of French. My grandfather, Morris Kanner, served on the battlefields of France during World War I; when I was a child, we discussed French literature. His daughter, Violette, won an award from the French government for her performances in French plays. With her as my mother—and Stewart Holland, with an academician's knowledge of French, as my father—how could I not share a love of the language?

My present-day colleague, Sylvie Deborde, was very helpful as I worked my way through this book. A postdoctoral fellow at the New Jersey Medical School and a native of Saint Michel Mont Mercure in the Vendée region of France, she patiently endured a bombardment of esoteric French idioms and references and unstintingly shared her knowledge. Other friends and colleagues made useful suggestions, too: Henri Tissot, Mike

Grant, Rachel Woodward, and Elizabeth Nesbitt. Last but not least, my wife, Jean Donahue, read the manuscript, and our discussions resulted in numerous useful improvements. Thank you all, and I hope that I have done the book justice. I don't agree with every remark they make, but I do think Georges Charpak and Henri Broch have some important and thought-provoking things to say, in any language.

Prologue: Sorcerers and Scientists

Two things are infinite, the universe and human stupidity.
But I'm not so sure about the universe.

—Albert Einstein

The Early Scientists: Sorcerers, High Priests, and Astrologers

Let's be clear that our aim is not to heap contempt upon sorcerers! We all start out bewitched, amazed, and frightened by the astonishing world into which we have been placed by fate. We learn to acquaint ourselves with it, to defend ourselves from it, and to understand it by forming beliefs, religions, philosophies, and sciences.

Ancient wizards—together with alchemists, astrologers, astronomers, and all those who sought to solve nature's mysteries—were the precursors of scientists, who discovered and modeled the world in which we live. They insatiably explored the unknown, with the aim of forming a coherent vision of the material and living aspects of the universe. These ancient seekers of knowledge were by no means isolated hermits. They were often allied with high priests, whose overarching ambition was to draw back the curtains from the most mysterious recesses of human nature so they could understand human destiny. For those who toiled making observations, the priests fulfilled the same role as theoretical physicists do for our experimenters today. Our present-day seekers are obsessed with revealing nature's secrets as they work with ever-more-powerful microscopes that provide glimpses of single atoms, giant accelerators that re-create the conditions of the big bang for fleeting instants, and telescopes that gather specks of light emitted 1.5 billion years ago from the edges of the universe.

Religion played an enormous role in the expansion of science but also in recurrent attempts to kill it off. Religion often put a brake on science's development and fiercely opposed anything that called dogma into question. From the moment when astronomers drove Earth from the center of the universe (and thereby from the focus of creation), the church persecuted them as common heretics—starting with Copernicus, condemning Bruno to the stake, silencing Galileo, and forcing Descartes into exile. It took centuries of political and social upheaval to change the relationship between certain religions and science. Just as in today's world, the illumination offered by science was considered beneficial by some, while others sought to snuff it out, and this led to conflict. The political battle at that early stage was, inevitably, won by science's opponents, given that the scientists' power was only that of a nascent movement.

The Human Soul Is Capable of Repentance

The recent apology by the Catholic Church and the rehabilitation of Galileo are signs of acceptance of the immense place occupied by science today in our perception of everyday reality. This acceptance, however, is not by any means shared by all sects. It is easy to find powerful fundamentalist groups in every religion, groups firmly grounded in the immutable truth of their dogmas, for whom science offers only the appearance of truth and hides, in reality, the devil.

We saw the opposite extreme in the nineteenth century, characterized by the flowering of what was really a religion of science itself, a naive faith in its beneficent omnipotence and in its ability to answer, someday, all the questions that trouble the human spirit. At that time, such a view stimulated the development of numerous groups that dreamed of reforming a society plagued by unacceptable flaws caused by the Industrial Revolution. This "religion" tainted the founding fathers of the Marxist revolutionary movement, in particular. On the time scale of human history, this outburst of ignorance seems risible: it affected the ideologues, who fancied themselves the most progressive people on the planet, for less than a century. But this historical experience reveals how easily people, even the most educated, can let themselves get carried away by aberrant beliefs.

After the collapse of the Soviet Union, most of its leaders recognized their error, as did most of the people involved (for a variety of sometimes noble and idealistic reasons) in the great human adventure of attempting to institute socialism. They came to understand that having a political party direct science was the same as having a church do so.

But that did not put an end to such aberrations. A powerful sect like the Church of Scientology still builds a religion around the pseudoscientific writings of its founder. The falsehood of these writings is so patent that they'd be touching coming from the pen of a ten-year-old. Everywhere there is a torrent of words, words, and more words borrowed from elementary science books—all devoid of meaning. This short excerpt from one of the founder's books will illustrate:

> *The Question Mark:* All this leaves us with an enormous question mark. Whether radiation travels around the world is beside the point. It's the question mark that flows around the world. And the question mark, if there, is the radiation itself.
> *The Effects of Radiation on the Human Body:* At what point is radiation harmful to the human body? No one knows, but we can state the following: a wall fifteen feet thick can't stop a gamma ray. On the other hand, a body can. Which leads us to pose this medical question, of the greatest significance: why can gamma rays go through walls but not through the body? Clearly, a body is less dense than a wall. As we do not find an answer in the material domain, we must therefore enter the mental domain. [L. Ron Hubbard, *All about Radiation,* 1967]

How can such stupid things influence anyone of even average education?

Science Turns the World Upside Down and Can Endanger Life Itself

The extraordinary success of societies that make unfettered use of the potential of modern science excites emotions of admiration mixed with fear that can result in hostility. This hostility is nourished by an apocalyptic vision—alas, sometimes justified!—of some of the consequences of scientific progress.

How can we accept climate change, occurring in the course of a century, that used to take ten thousand years?

How can weapons capable of wiping out life on Earth be deployed, when measures for their control depend upon political formulations that have failed in the past?

How can we believe that there will be billions of additional people in the near future, as demographers predict, who will be trapped in immense regions of poverty, and that at the same time we will we enjoy our industrial civilization, full of consumer goods, in perfect tranquillity?

It is the unbridled pace of scientific discovery and its applications that has unleashed opposition. It's no coincidence that the effects of science and the ability of people to manage them are being called into question now. Maybe this is a healthy reaction. It is coming forward at a point when spiritual groups, intertwined with our societies throughout the planet, hold divergent positions on how to participate in the inevitable advancement of science while neutralizing the noxious effects of its intrusion into societies that are not prepared for it. Our instincts originate in large measure in our genetic inheritance, optimized for cave dwellers tens of thousands of years ago after a two-billion-year evolution of life on a planet four or five billion years old.

To get an unsettling, although pertinent, idea of the extraordinary challenges facing us in the century ahead, imagine what would go through the mind of an alien from another planet in our galaxy who had been observing human activity for two centuries. He would be astonished at how we have depleted our fossil fuels in such a short time. These deposits of coal, oil, and gas were built up over tens of millions of years in a complex cycle of accumulation of animal and plant residues. Now, three billion additional people claiming their share of the energy needed for a decent life (as defined by the example we have set for them) are not the only problem we need to worry about. The climate changes caused by irresponsible exploitation of resources can, on the modest time scale of this century, produce catastrophes that will throw hundreds of millions of people onto the shores of those countries that have been spared.

The people of Asia, Africa, and the Americas who were slow to industrialize got crushed, and their treatment warns us of what may happen if an awareness of the fundamental unity linking the destinies of all people on the planet doesn't become the driving force behind international relations. With its persistent refusal to yield to worldwide control of greenhouse gases, the United States offers a perfect example of the blind selfishness surrounding the exploitation of natural resources today. To turn away from such behavior, people must be well aware of consequences. The paradox is that only science itself can illuminate this reality and be the indispensable tool for undoing the damaging effects of science.

We live in a world in which the level of scientific development makes possible the manufacture of easy-to-transport, inexpensive weapons of mass destruction by groups with little territory but sufficient money at their disposal. This makes immense tragedies inevitable if the level of compassion of the leaders of the industrialized world toward other nations remains that of their predecessors, the practitioners of slavery and colonialism.

If we seem to champion moral values like the spirit of compassion and solidarity, it's because we believe that the extraordinary spiritual richness of humankind is not manifest solely in scientific progress. Art, philosophy, the social sciences—all honor humanity. But why are writers, poets, and politicians sometimes as ignorant of scientific matters as might be the sorcerer of a lost tribe in an uncharted forest or a fundamentalist religious leader?

We've always admired the "science for poets" courses given in some of America's best universities, generally by acclaimed senior experts. And no government officials can avoid some knowledge of science at this level, if they are to decide on anything other than administrative matters.

We do not expect that this book will change the course of human history. We hope only that, in commenting upon some commonplace experiences of sorcery, among friends, with a smile, we will show how some modern-day sorcerers deceive their poor audiences! The wizards are often well established and may even win university positions. In no case do we want to impose a single way of thinking, not even the scientific one. We are agitating, on the contrary, for doubt, skepticism, and curiosity. But of course, we do have the greatest respect for true conjurors, illusionists who, for our greater happiness and that of our little children, perform tricks that leave our mouths agape!

NASA, Apollo 11

Our Amazing Ancestors, the Cavemen

Here we see Earth photographed from the Moon. It is a token of the power of science and technology that, thanks to satellites, space stations, and probes sailing through space for dozens of years, a rich harvest of observations lead to the discovery of new phenomena that are far from fully understood. Discoveries in space will lead to a confirmation or rejection of hypotheses, like the big bang, that bear witness to the creativity of human beings, creatures who can conceive of mechanisms involving time scales and distances so vastly different from those they experience. An examination of the earth and its environment, however, shows that the intrusion of science and its consequences may threaten the existence of the very societies that give rise to the amazing creativity of science.

Life developed on Earth because of an extremely rare confluence of characteristics, including a favorable atmosphere (which appeared approximately three billion years after the condensation of the dust of dead stars) and the planet's mild temperature. This latter characteristic is to some extent due to internal heating produced by a sphere of molten metal seven thousand kilometers in diameter, nestled in the center of the planet, which draws its heat from the radioactive decay of the original stardust particles that formed the earth. It is mainly solar radiation, however, that provides Earth with the heat necessary for life.

No other planet in the solar system enjoys these advantages. What is worrisome is that, at the start of the new millennium, the amount of heat that escapes from the molten core and is radiated out to intergalactic space is about the same as that generated by human activity, which draws energy willy-nilly from fossil fuel sources that will be depleted a few centuries from now. The comparison would be of little importance except for global warming, which threatens the survival of a portion of humanity in the first century of this new millennium. It's clear that human societies must use their intelligence to face up to this threat, and to this end they must draw upon the powerful resources offered by science.

We have mentioned the genetic capital that comes to us from our ancestors, the cavemen, and that has no doubt retained its original force. The lifestyle of the majority of their descendants has been generally and considerably modified, with unbridled globalization rooting out the most isolated tribes and dragging them into the consumerist norms of the standard Earthling. It would be a mistake to believe that how people think has undergone profound changes since caveman days, especially where spontaneous reactions to unforeseen events are concerned. Those excluded from scientific and technological thought processes have reactions to events that are the same as those of their cave-dwelling ancestors (who did, after all, leave a magnificent heritage from which we draw the best values of our civilizations).

The harsh struggles for life and the need to invent thousands of ways to overcome the difficulties of existence led over the course of millions of years to an end result that gives us plenty of reason to be amazed. But we are also confronted by the fact that the human species, because of the science it has developed, could destroy itself by unrestrained recourse to behaviors that were always reprehensible but at least did not previously pose the risk of a planetary apocalypse. The consequent need to invent a new way of acting for society will require that the bulk of human societies master scientific reasoning, and this runs against the innate tendencies of people to protect the material and spiritual niches that assure their own survival at a given time. These tendencies are manifest with a force and virulence that are all too human and take very diverse forms that we will attempt to demystify: superstition, astrology, the paranormal, clever fakery, and the like.

We must be clear about this: neither of the authors considers himself such a repository of wisdom as to justify giving his fellow travelers an opinion on the big choices in their lives, particularly in spiritual matters. Science can never pretend to have understood what makes people tick or the meaning of life, nor can it even prove that it will ever do so. We will attempt to take as our own the rather desolate profession of faith of Stig Dagerman (*Notre besoin de consolation est impossible à rassasier* [Arles: Acte-Sud, 1952], 1):

> I am lacking in faith and therefore cannot be happy, for a man who risks fearing that his life is a senseless path toward certain death cannot be happy. My upbringing provided me with neither a God nor a fixed place on the earth from which I could attract the attention of a God. Nor did I inherit the well-disguised fury of the skeptic, nor rationalist tricks, nor the passionate lack of guile of the atheist. Thus, I do not dare to throw stones at those who believe in things that inspire me with doubt, nor at those who cultivate their doubt as if that doesn't also engulf them in darkness. The stone is for me, for I am positively certain of one thing: the human being's need for consolation is impossible to satisfy.

It would be disastrous if that need for consolation resulted in extreme vulnerability to the siren songs of the merchants of deception who cross our paths.

People possess a great treasure: their free will, which creates the possibility of choice. Each person's brain holds more connections—billions and billions of times more connections—than the most powerful computers made today. Moreover, the individual's life, connected with that of billions of other people, is open to infinite possibilities for social relations over great time spans. It would be pure madness if the promise contained within such infinite potential got destroyed in primitive conflicts to which science has given devastating force. We think that the mastery of the elements of scientific culture by the greatest possible number of people will turn out to be even more decisive in the future than speaking, writing, and money have been in the past.

Our aim is to offer readers the chance to work on some exercises, for enjoyment and edification, that will allow them to master techniques used by those who make the decisions needed for the human race to adapt to the changes caused by our marvelous creativity. By learning to fool others, you will be better prepared to judge the blandishments of the merchants of deception, who try to persuade you of their special and extraordinary knowledge, whether it's in the area of health, love, or politics.

Remain scientists—and become sorcerers!

Earthlings, Radioactive from Birth

Among the ashes of burnt-out stars whose condensation created Earth, there are certain heavy radioactive elements, including uranium and the like, but potassium as well. You find potassium in all living tissues. It is necessary for life. Its absence leads to serious, and often lethal, illnesses.

In her ignorance, Mother Nature neglected the fact that a hundredth part of this essential building block would be radioactive and might terrify an anxious ecologist. Its half-life is 1.3 billion years, so this enduring radioactivity is easily measurable, thanks to the wonderful sensitivity of particle detectors. Accordingly, we know that the body of a human adult has about six thousand potassium atoms disintegrating per second. This radioactive decay gives off electrons, and high-energy gamma rays in particular, which leave the body and irradiate any innocent and no doubt unsuspecting person sharing a bed. Scientists and citizens' groups concerned with radioactivity tell us that six thousand becquerels' worth of radiation is in each body because of potassium. Of course, that's not our only source of radiation. Our natural environment, the earth and the sky, generously donate at least twenty times that.

What's amazing, though, is that a shrill alarm goes off when detectors pick up contamination one-tenth or one-hundredth of what we get from naturally occurring potassium. Such contamination is a gold mine for businesses and a springboard that certain groups can use to obtain political power. Maybe you've seen on television engineers who are deeply concerned about their fellow citizens' problems and who run independent laboratories to measure radiation. Maybe you've seen them crying wolf on prime time shows, announcing an important new crisis: Fiberglass is radioactive; powerful industries are responsible! Even the beach providing the sand from which the fiberglass was made is radioactive! Sure, it's true, but beach sand is simply the product of the erosion of naturally radioactive rocky outcroppings, transported by water currents to the seashore. You've probably seen announcements of costly studies, with tabs running into the millions, being launched to analyze the toxic effects of depleted uranium used in antitank shells in the Gulf War. It's obvious that there's no toxic effect due to the radioactivity in those shells; they emit less radioactivity than you would breathe in from radon if you were lying in the grass sniffing the flowers of the field. Radon is a naturally occurring radioactive gas produced by the radioactive decay of the uranium present throughout the earth's crust and in many houses. The adherence to an ideologically motivated prejudice leads inevitably to astonishing misrepresentations of reality.

We mention all this because it is legitimate and ultimately reassuring that citizens worry about the relative harmfulness of sources of energy, including nuclear power, that are consumed by their society. But it's not reassuring to see how more or less well-intentioned leaders can exploit the ignorance and fears of their citizenry to get them to take decisions that may be catastrophic for the planet. The leaders' intentions are not generally corrupt. They aren't really bankrolled by economic forces linked to various sources of energy, but they never hesitate to exploit ignorance as a powerful political tool.

It's very sad that, a hundred years after the discovery of radioactivity by Henri Becquerel and Pierre and Marie Curie, the vast majority of citizens, even educated citizens, have never handled a Geiger counter. That's like never having used a ruler to measure something. People have learned to take note of one atom decaying, when there are billions and billions in one drop of dew. They could have learned that, in a drop of water from their local river, there are millions and millions of atoms of arsenic—but that's not why they shouldn't drink the river water. People would be fascinated by the syncopated rhythm of the Geiger counter's music, a rhythm that would help them to understand the secrets of the laws of chance that are an essential element of our understanding of the world. And the subtleties of the laws of chance are among the tools we use to snuff out the power of those who would deceive, as we will discover later.

We Sail Away on an Ocean of Ignorance

Now we are going to set sail amid islands of ignorance, which are perfectly well charted in any good atlas of superstition—and which are objects of veneration and pious pilgrimages nonetheless. They are so numerous that we must content ourselves with a few well-known examples:

- How can we exploit a physical phenomenon little known by the public to pretend that we possess extraordinary powers, such as being able to bend keys in the pockets of gullible individuals?
- How can we levitate?
- How can we free ourselves at last from the shortage of water that threatens humanity?
- How can we become astrologers with exactly the same chances of accurate predictions as the best in the profession?
- How can we become telepathic, freeing ourselves from the limitations of space and time?

Probability, the study of chance phenomena, is not ordinarily very familiar. Lack of knowledge of probability leads to error even by distinguished professionals in such fields as medicine, physics, and, of course, the social sciences. As a systematic field of mathematics, probability theory did not appear until initial work by Pascal and Huygens in the seventeenth century. In this book we do our best to show how lack of knowledge of probability is one of the most pernicious sources of superstition and deception. The teaching of its fundamentals should have an important place, like that of arithmetic, in all mathematics curricula, at every level.

Debunked!

Chapter 1

The First Steps in the Initiation

Your Initiation Begins

Your powers are finally going to be revealed to you, and you are going to learn how to control nature. You will learn how to

- describe accurately the personalities of people you don't know
- have ecstatic visions that are entirely under your own control
- get on the same wavelength as another brain and practice telepathy
- levitate, pierce your tongue, order your heart to stop
- walk unharmed on hot coals
- bend metal just by concentrating your mind!

Truth Is Drawn from a Well

"The proof that astrology works, and that it works well, is that my horoscope has accurately predicted things that actually happened to me." How many times have we heard such remarks? How many personal experiences of this type get presented as evidence of the validity of astrology?

Well, let's be clear on the subject: Yes, horoscopes work—they work well, in fact. But the validity of the horoscope does not imply the validity of astrology. Many people are convinced of the validity of astrology because their horoscope "works." These people believe that the occurrence of predicted events, which they have witnessed, justifies the validity that they ascribe to the "Science of Signs." They are especially convinced that their horoscope gives them a solid foundation for an understanding of themselves and provides guidance for how to act and on their destiny.

For such people, their horoscopes are meaningful, but in truth, the horoscopes take on meaning *from* the believers, not *for* the believers. It is difficult to get this point across, since it runs contrary to personal experience—"You can't say it's not true, because it's happened to me." The individual who reads his horoscope is convinced that he is dealing with *his* horoscope, that that horoscope is destined for him, and that it was created specifically for him by a supernatural force. There's no recognition that the satisfaction of the client is a source of feedback, adding to the credibility that the fortune-teller can claim for himself and his "science" and consequently for his motives and his effect on the client.

A Convincing Demonstration

Twenty years ago, during a class on paranormal phenomena and the occult, one of us asked the students to write the following information on a piece of paper: their first and last names; the date, time, and place of their birth; and the theme of their latest dream. All of this was handwritten. The request implied that some kind of star-based calculation would be made using the birth data or that a handwriting analysis would be made based on the written material, in either case possibly augmented by interpretation of the most recent dream.

A week later, each student received an *individualized* description of his or her personality followed by the question, "How good is this description of your personality?" The concordance of the description with the student's real personality (or self-perception of it, anyway) was rated as excellent, good, fair, poor, bad, or none. Overall, 69% of the students judged the description of their personality to be excellent, good, or fairly good.

This result is especially convincing because I had been introduced to the students as a "scientific demystifier"—a pretty lousy introduction, by the way—and not as an astrologer or other supernatural leader, which would have certainly enhanced the believability of the personality descriptions and thus the percentage of "successful matches." It's also a particularly good demonstration when you consider that, when we asked one of the students to read aloud the personality description that we had prepared for

- You need others to like and admire you, yet you are apt to be critical of yourself.
- Although you have certain character flaws, you are generally able to make up for them.
- You possess substantial untapped potential that you haven't exploited for your own benefit.
- Some of your aspirations tend to be pretty unrealistic.
- You are outwardly disciplined and you display self-control but inside you tend to worry and be insecure.
- Sometimes you have serious doubts as to whether you have made the right decision.
- You prefer a bit of change and variety and are annoyed when you encounter restrictions or limitations.
- Sometimes you are extroverted, affable, and sociable, while at other times you are introverted, cautious, and reserved.
- You are proud of being an independent thinker and don't accept the statements of others without satisfactory proof.
- You find it unwise to reveal too much about yourself to others.

him, the others all thought it was theirs. They couldn't have been more similar—in fact, the much-touted "individualized" descriptions were prepared before the students' data were collected, and they were *exactly identical* for all the students! This was a simple but enlightening demonstration of one of the many "effects" that occur so often when it comes to "paranormal" phenomena. If you ever want to try this illuminating experience, use the model personality description above; simply add a name to complete the "personalization."

Similar descriptions were used and tested for the first time by the psychologist Bertram Forer, who based his wording on an astrology book. The effectiveness of such wording in a real-life setting demonstrates clearly the power of what some have called the "well effect."

The Well Effect

The well effect can be summed up as follows: The vaguer a statement is, the more numerous are the people who will recognize themselves in it—

and the more completely the description will be seen to "fit" them. The descriptive statements may be profound, all right, but they are only deep in the sense that a well is deep—deeply hollowed out, that is, empty. In fact, experience has shown that vague, general statements are *more* convincing than specific descriptions made by professional psychologists because of what sociologists might call the Barnum effect. Barnum's circus shows were constructed so that there was something for everyone so that everyone would find their way there, which created the shows' success. Moreover, studies have shown that, when it comes to analysis of serious personal problems, "yes" and "no" answers picked in advance completely at random are perceived as very encouraging answers to specific questions by the people posing them! (You can read more about this phenomenon in Ray Hyman's "Cold Reading: How to Convince Strangers That You Know All about Them," in *Paranormal Borderlands of Science* [Buffalo: Prometheus Books, 1981].)

The well effect explains, in large measure, the success of horoscopes. "In some ways, you are one of the stronger people": this is a statement that is empty and devoid of meaning, yet it will be accepted as fundamentally *true* in a horoscope, as each reader interprets it in the context that gives it meaning for him. The reader will think, "My knowledge of foreign languages is strong," "I have great ability when it comes to fixing things," "My muscles are strong," or any of so many other possibilities. And this is without even taking into account that the elementary principles that can improve the receptivity of the readers. For example, don't tell people what you know (or think you know) to be true about them, just tell them what they *wish* were true about them.

Of course, astrologers count on the fact that the public quickly forgets the predictions. The supermarket tabloids print dramatic predictions at the end of every year, often involving such events as a presidential assassination or the second coming of Christ in the following year or two, and people have been reading similar predictions for decades. In France, a prediction concerning a former prime minister is little remembered: "Despite a generally positive picture for the year [1993], the first fifteen days of January and of September as well may pose serious problems for Pierre Bérégovoy." For this prediction we are indebted to a famous astrologer, Elizabeth Teissier, writing in *Your Horoscope for 1993*. For those unacquainted with subsequent events, Pierre Bérégovoy killed himself on May 1, 1993, by a gunshot wound to the head. In the United States, thousands died in the World Trade Center disaster of September 11, 2001. The horoscopes of any number of the victims contain predictions supposedly useful for guiding behavior that day—"a good day for a fresh start in your love life"— and the next day! Individual deaths on that day were not predicted, let alone a mass disaster; if astrology can't unambiguously predict an event of that magnitude in your life—and still makes further predictions for the next day—what good is it? Yet no one remembers that predictions of good days and bad days for the rest of the year were cast for these victims.

Clearly, astrologers make full use of the well effect. A prediction from Elizabeth Teissier again: "Populations in the world will suffer violence in the following month, given that Venus and Pluto . . . " But they also use other tricks to lend credence to their pronouncements, always centered on the big three—Love, Money, and Health. And intellectual honesty does not have to be the foremost characteristic in the star charts that predict the traits of the astrologers themselves. Skill and craftiness are often plainly evident. For example, Darah was one of the four sons of a great Moghul emperor, and he was obsessed by astrology. A soothsayer predicted that he would stake his own life on his prediction that Darah would be crowned the heir to the emperor. When someone expressed astonishment at the temerity of the prediction, the astrologer said, "One of two things will happen: either Darah will ascend the throne and my fortune is assured, or he will lose his bid and be assassinated, and then I will have nothing further to fear from him." Beyond the well effect, astrologers don't hesitate to have two irons in the fire or, more subtly, one iron with two different sides.

Let's not leave out that astrologers also count on the fact that "to err is human" but constant fallibility is not. Neither perpetual fallibility nor perpetual infallibility is characteristic of anybody. People readily accept one side of this and disregard a claim of infallibility because errors inevitably appear. But people are a lot less aware that a few instances of "getting it right" are inevitable. The contrary—always being wrong—would be extraordinary. Even an astrologer will sometimes make predictions that turn out to be true. So, predict merrily along—some of your statements will always pan out.

Astrology in a Vacuum

Astrologers actually know very little about what goes on in the skies. In *Your Horoscope for 1993,* Elizabeth Teissier asked, "How does such a horoscope [for a group readership] work, and how can it be justified? How can it be conceivable that a Capricorn born January 9, 1960, should be under the same planetary influences as another born January 9, 1924, for example? Here is the answer to such questions: In its apparent travels around Earth, *the Sun ends up in the same place in the sky on the same date each year"* (italics added). This is totally false! On two identical dates in different years, the Sun is *not* in the same spot in the sky at all.

On a given date in different years, our planet is not at the same point in its orbit around the Sun. As we will explain more fully later, the phenomenon of the precession of the equinoxes causes a shift in position. To put it in numbers, as an approximation, we can say that there is a difference of about twenty-two thousand miles between Earth's location on any specific date in two successive years, which is a distance of about three times the diameter of our planet. Thus, contrary to the astrologer's statements about the same planetary influences on January 9th of 1960 and 1924, the Earth would not occupy the same place in its orbit around the Sun, at all. Between those two dates there would be a shift of about 780,000 miles!

For the most part, astrological soothsayers work with what is called the "tropical zodiac." This is based on the sun and has nothing to do with the stars, which are the basis for the "sidereal zodiac." When astrologers put together an astral analysis, they generally use the twelve signs of the tropical zodiac, defined as the twelve equal rectangles into which the celestial sphere is divided. The starting point for this division is called "gamma," which is the intersection of the ecliptic with the celestial equator, corresponding to the spring equinox.

At one time, a bit before the birth of Christ, calculations based on the tropical zodiac would have been about the same as those based on the stars and constellations that originally determined the characteristics of the various signs of the sidereal zodiac. This is no longer the case because the precession of the equinoxes has mixed things up by displacing the gamma point with respect to the starry background of the celestial sphere. This displacement has dragged along the astrological signs, which nowadays don't correspond at all to their original stars. Today's "tropically based" astrologers just blindly apply rectangular sign-zones, empty boxes that have nothing to do with anything and are devoid of any consistency or correspondence with the stars. If you want to know your *real* zodiac birth sign, rather than the one fobbed off on you by the usual (tropical) astrologers, we'd advise you *not* to consult the hottest astrologer who's become a media darling—investigate *Astronomic Zodiac* instead. This resource is found on the Internet site www.book-e-book.com. Your true sun sign is simply the location of the Sun in the celestial sphere, as seen from Earth at the instant of your birth. This can be calculated rigorously from an astronomical point of view. But most astrologers, relying on the vacant rectangles of the tropical zodiac, really don't practice astrology at all, but rather something we've got to call the study of emptiness—voidology, or astrology in a vacuum.

The Navel Gazers

Horoscopes are popular today because ours is a narcissistic civilization. Science makes only global or collective forecasts, while lots of people are interested only in their own personal destiny. For most people, it's not hard to choose between the distant scientists who speak of generalities and the accessible astrologer who speaks exclusively to the individual about himself. The aura of exclusivity and uniqueness is certainly enhanced by the astrologer's request for the complete details of the client's birth. Exact spot, date, hour, and minute—all pertain to a single person, all pertain to *me*— so it *must* be that there's a good correspondence between the results of the astrologer's study and *my* own personality.

Our observations and perceptions depend in part upon what we are thinking at the moment we observe something. Our deepest desires and motivations, modified by our past experiences, are reinforced, whether consciously or not, by selection bias. Selection bias is a well-documented psychological principle. It means that we choose our magazines, newspapers, radio stations, television shows—all of our information sources—in such a way that our opinions are, for the most part, reinforced rather than challenged. And if, despite all that, we receive challenging information, we can always use subjective validation. Subjective validation is a psychological principle that allows us to absorb *incorrectly* any information that is contrary to our preferences and to interpret it in a different light. Subjective validation causes two events to be perceived as linked, when they are *not,* simply because a desire, a hypothesis, or a belief requires such a linkage. In the case of astrology, events are perceived as linked simply because the horoscope says the linkage exists. This perception of linkage, in turn, induces superstitious behavior founded upon the belief that one's own actions determine the course of events, even when this is not so in reality.

If the influence of the planets over our destiny is completely nil, however, it doesn't follow that the horoscope is without effect. And the well effect, in particular, allows us to see why the horoscope holds such a sway over so many people.

The Precession of the Equinoxes

The precession of the equinoxes was discovered by Hipparchus of Nicea in the second century B.C.E. Earth is not perfectly spherical but flattened a bit at the poles and bulging a bit at the equator. Gravitational effects of the Sun and the Moon on the equatorial bulge cause the axis of Earth's rotation to shift. That is, the axis—the line between the poles—moves around. It pivots, a bit like a spinning top, very slowly, taking about 25,790 years to go around completely. The picture is actually a bit more complicated, though. An additional phenomenon, called "nutation," creates a little wave around the main circular motion of the axis, with a period of about 18.6 years.

In about twelve thousand years, Earth's axis will point to a new "North Star," Vega (see figure on facing page), and today's "North Star," Polaris, will no longer mark the northerly direction.

The plane of the celestial equator—that is, the plane that includes Earth's equator—obviously follows this pivoting of Earth's axis. Therefore, its intersection with the ecliptic, which is the plane of Earth's orbit around

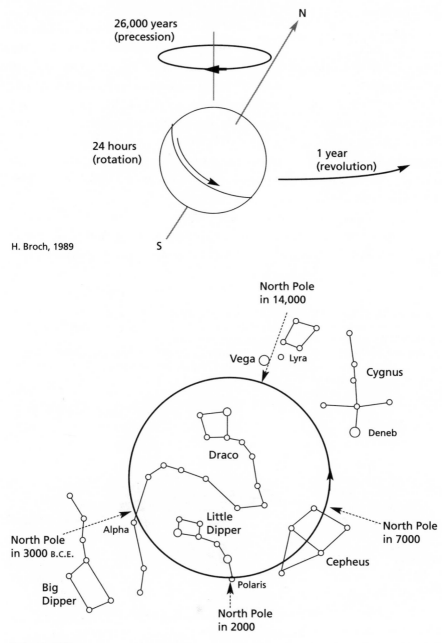

26,000 years
(precession)

N

24 hours
(rotation)

1 year
(revolution)

S

H. Broch, 1989

North Pole
in 14,000

Vega ○ Lyra

Cygnus

Deneb

Draco

Little
Dipper

North Pole
in 3000 B.C.E.

Alpha

North Pole
in 7000

Cepheus

Big
Dipper

Polaris

North Pole
in 2000

H. Broch, 1989

the Sun, must also move. But the location of that intersection determines the gamma point, the spring equinox. (Remember that the spring and autumn equinoxes correspond to the two positions of Earth when the line from the Sun to Earth is perpendicular to the axis of Earth's rotation. That's why, at the equinoxes, day and night are of equal length all over the world.)

The problem is that the spring equinox serves as the reference point for the tropical zodiac. In other words, the gamma point moves slowly but surely through the celestial sphere and pulls along with it the signs of the tropical astrologers, who consequently are not working from the original constellations at all and who continue to be farther and farther away from them (see figure below).

Here's just one example among many: Those born when the Sun (as seen from Earth) is in the constellation Leo are said to be "courageous, proud, and dominating." But if that were true in Hipparchus's day, two thousand

The position of Earth in its orbit at the spring equinox today and its position at the same equinox in the year 4600. Twenty-six hundred years from now, Earth will have completed a tenth of the complete cycle of precession. At the spring equinox today, the Sun as seen from Earth is in the constellation Pisces. In 4600 at the same equinox date (around March 21), the Sun will be in the constellation Capricorn!

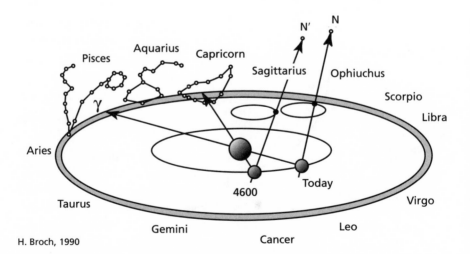

H. Broch, 1990

years ago, it's really difficult to understand what that sign has to do with us today. People born at the end of July are "Leos" according to the astrologers, but today's Sun is not then in Leo at all but in Cancer.

Interlude

The well effect is not limited to astrology. One can easily find many applications of this principle in every aspect of life; the box on the next page gives an example from the field of politics. The original text, little altered here, is from our friend Jacques Poustis, who reintroduced this type of practical exercise by publishing it in his magazine, *The Fairy Said So,* in January 1998.

Put a statement together by beginning with the box at the upper left and then selecting one box at random from each of the four columns numbered 2 to 4. Repeat this process, always reading the chosen phrases in numerical sequence (1 to 4). Be sure to use appropriate intonation and forceful conviction when reading, as these are the indispensable ingredients of persuasion.

Personal Experience

"Personal experience" commonly plays a role in all kinds of beliefs in the paranormal. Every rational argument comes up against a reply like "You can't tell me that it's not so because it actually happened to *me.*" How can people be convinced that their own experience does not constitute proof? If someone swears to you that yesterday was his lucky day because he won a lottery jackpot, it would be hard for you to argue with him that he hadn't considered that day special in advance of hearing the good news and that any day would be labeled "lucky" in the wake of good tidings.

Personal experience isn't proof, for a variety of reasons, but primarily because what we recall of our experience is often quite subjective. Even an apparently concrete experience is subjectively reported—in other words, *falsely* recalled. Personal experience has an effect on our beliefs about events that have already happened—memories that have been retained are affected. Experience is by definition something that happened in the past;

Ladies and Gentlemen:	the present situation	must be part of the overall finalization	of a process leading toward more equality.
I remain deeply convinced that	the exclusion that some of you feel	demands a much more effective pursuit	of a future oriented toward more progress and more justice.
You should know that I will exert myself so that it becomes understood that	the seriousness of the problems of daily life	motivates me as a citizen and requires all of us to be in the forefront	of a social restructuring in which each individual will finally regain his dignity.
It is with full knowledge of the facts that I can tell you today that	the fierce desire to get our country through this crisis	inevitably leads to an urgent requirement	of an uncompromising respect for us as individuals.
I tell you here today of my unfailing determination to argue forcefully that	the compelling effort to improve the precarious situation of the excluded	strengthens my unshakeable desire to go in the direction	of a plan that really meets the legitimate needs of everybody.
I hardly need to remind you that I have long supported the idea that	the special circumstances owing to our unique history	must lead us to the truly necessary choice	of rapid solutions to deal with our great social priorities.
It is with a clear mind and strong conviction that I tell you that	the more than legitimate social aspirations of all people	requires a focus on the people at the bottom during the unfolding	of a more human, charitable, and just program.
Surely you will not contradict me, dear fellow citizens, if I tell you that	the need to reply to your everyday concerns, whether you are young or old	engenders a mission that I hold in the highest regard: the development	of a project that can bring true hope, especially for the least fortunate among us.

thus, it is something that one remembers—it's supposed to guide us and help us interpret and make sense of the world. But it doesn't really constitute an explanation. It doesn't provide a proof. What's really going on is not simple at all.

What Do You Remember?

Now let's put your *visual memory* to work. Many people have a good memory for images, a sort of mental photography of an event at a given instant, and we call this "visual memory." These people have the ability to retain what they have seen, pretty exactly.

Perhaps you are one of them. Concentrate for an instant and recall the situation when you had breakfast this morning: recall what room you were in, what exact location; what you drank, what you ate; what you were wearing; and so forth. In short, recall a picture of the scene. Don't turn the page until you really give it a try.

Do you recall an image like this one?

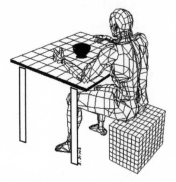

An image in which you are seated while eating breakfast? Perhaps you imagined yourself from the back and a little above and to the side, as shown here. You have just "seen" a scene in which you are a central figure.

But let's think about this: you must really have seen only your two hands and a bowl! Rare are those who recall the scene through their own eyes, as such a picture, with solely those elements. Most people recollect the scene as we described it above, in terms of a picture that certainly never struck their retina.

In fact, the process of creating a memory requires its active construction when we wish to remember something, and that construction is by its very nature a process of re-creating and elaborating. That's why, when someone tells us about a personal experience that they actually lived through, which proves beyond doubt the existence of some paranormal phenomenon, we don't doubt their sincerity, but we must consider their statements extremely cautiously.

What Do You See?

Let's stick with visual examples for a moment. Did you know that the eye's abilities have specific limitations? "Seeing is believing," people say, but that's not true, as we shall see.

Stare for thirty seconds at the picture on the following page. Focus on the little dot between the two hearts. Without moving your eyes and without blinking (or keeping blinks to a minimum), count slowly to thirty in your head. Then—and only then—turn the page and continue reading.

From Jacques Poustis, "Fou! . . . mais logique," *Science et pseudosciences* (Association française pour information scientifique—AFIS), August 2000, no. 243, pp. 2–5.

Stop! Stare at this blank page before proceeding!

What you just did must have made you "see" a picture like this:

It's a picture of Christ (or Karl Marx, or your bearded friend or colleague, depending on your presuppositions) that appeared on the blank page and moved around a little. If you looked somewhere other than at the blank page, on a fairly uniform surface like a nearby wall, you would have seen the same image but on a much smaller scale.

The apparition appears because of two phenomena. The first is *retinal persistence,* which causes even a floating image to last a certain duration of time, about a fifteenth of a second. Retinal persistence allows us to see a series of discontinuous projected fixed pictures as a continuous motion, like when we watch movies that are stored on film.

There's a second explanatory phenomenon. The image-processing cells in your retina are composed of rods (at the edges, for low-brightness viewing) and cones (in the center, for brightly lit scenes). If you stare at an image without moving your eyes for at least a couple dozen seconds, these retinal cells get saturated and can't transmit changes in light. If an original image is abruptly removed and your gaze transferred to a uniform surface, the retinal surface can transmit colors normally except in certain places where cells saturated by the previous picture can't provide optimal transmission for the colors present. Consequently, we perceive an image that is complementary in form and color to the original image.

Since this complementary image is actually in the eye, no matter where you look you will still see it. The image is on the retina independently of what's outside, so the farther away you focus your gaze, the bigger the phantom image seems, because the angle it occupies within the field of vision always remains the same. As your gaze moves from the blank page in this book to a more distant wall, the angular dimension of the phantom stays in proportion, and it can only get bigger.

To clarify, suppose you view a 2-inch image on a page 12 inches from your eye. The image will occupy an angle of about 10 degrees of arc at the peak of an isosceles triangle having a base of 2 inches and a height of 12 inches. The phantom image formed by retinal persistence will have exactly the same "dimension" on your retina. Thus, if you look at a blank page 6 inches from your eye, half as far away as the original picture, the angular field covered by the phantom remains 10 degrees, so the afterimage will be half as big (1 inch). Similarly, if you look at a blank wall 10 feet away, 10 times as far away as the original drawing, you will see an image again occupying 10 degrees of your field of vision—a 20-inch phantom.

We chose a black-and-white test picture for simplicity, but the phenomenon works with equally great effect in color. So if you want to see a green alien, draw a picture of the extraterrestrial of your choice and color it violet with big white eyes. Focus on the picture for a minute before transferring your gaze to a white surface, like a wall or refrigerator, and watch your friendly alien appear—a green alien with big black eyes.

Practice Telepathy!

Imagine the following situation. You are with friends, happily settled in armchairs after a good meal, and the conversation (by happenstance) turns to paranormal phenomena. They are well-known phenomena, "well proven by numerous widespread experiences," which "science, always 'narrow,' with blinders on to avoid seeing, absolutely refuses to examine."

Telepathy, the power of brains to communicate directly with each other, comes up sooner or later, from some angle or other: "Haven't you ever received a phone call from someone at the precise instant when you were thinking about them?" Or, "I can tell when I walk down the street that someone is looking at me even though he is behind me and I don't look back at him." "And what about twins? Amazing, isn't it, that power that they have to feel just what the other is feeling, although they are miles apart."

That's when your spouse interjects: "By the way, honey, have you told your friends yet that you have a gift that's a bit unusual? That you became aware of it by chance? You can communicate with a colleague at a distance, even a very great distance, just by thinking!"

And you, in complete innocence, answer, "No, no, let's not mention that!" (It's important to refuse to talk about it. Reticence makes it a much more serious matter.) "I'm not a medium or anything of the kind," you continue. "I don't know how to explain this gift, and I don't even know how it works." (Modesty in your explanation is always profitable. Your friends can then show off their knowledge of the subject by declaring that scientist X of institute Y has proven that telepathy can be explained by principle Z.) "But all right, it's true; I have the gift of telepathy."

After such an opening it is certain that someone there will beg you to tell more about it and will insist on a demonstration to prove your claim. You then explain that, for this telepathy to occur, there must be two brains on the same wavelength and that you discovered such a neural concordance with a colleague after several experiences. This colleague, who is with his family right now a few miles away, is (say) Mr. Norris (you name him explicitly). You are able to concentrate on a playing card and, just by thinking about it, you can tell Mr. Norris which card it is. This works pretty often, not always, but pretty often. (It is important to be hesitant and to be restrained in your claim, short of 100% effectiveness.)

Then you yield to the friendly pressure to make an attempt then and there. "Okay, I'll try," you say, "but I can't guarantee anything. I'll have to call my colleague, and I don't even know if he's there. I think his phone number is . . . " After giving it you ask, "Do you people have a deck of cards?" Someone brings you a deck. You don't touch it; you ask someone else to shuffle it and have it cut again and again, and then you have the other people present decide among themselves who will draw the card at random. Finally, a card is drawn—the seven of clubs, for example.

Before getting ready and concentrating on the card, you take out an address book to double-check the telephone number, and you jot it down together with your colleague's name on a piece of paper. You give this to the person who's been chosen to make the phone call. The call will be made from a different room from the one you'll be in when you are mentally transmitting the image of the card, the seven of clubs, over a distance to your colleague.

The card is placed in front of you, and holding your head in your hands, you focus on it while taking deep breaths. Seconds, then minutes slowly pass. The people there have you in their sight the entire time. Ten minutes later, the person who made the call returns with undisguised astonishment: unbelievable though it may seem, Mr. Norris said—after much hesitating and receiving of hazy visions—that he perceived a seven of clubs! You are totally spent and out of breath. You hold your head saying, "Whew! That's really exhausting. I won't do that again for a long time. It's really hard and I think my brain is overheated."

All doubts melt away, quickly transformed into testimonials. "It's amazing! The cards were handled only by us. You told us his name in advance, and his phone number, too. It's not your house, so the room can't have been set up to trick us somehow—no hidden transmitter."

The person who made the phone call swears that the person who answered didn't squeeze information out of him. The caller simply asked for the appropriate person, explaining why, and gave no information of any kind that would have given any indication of the card chosen.

What happened was truly extraordinary. One or two people may still harbor some doubt but lack the power to make a reasoned argument. It's more a doubt in principle, coming from narrow-mindedness, of course. Could we see the demonstration one more time to be sure? Impossible to do it again—you already said, it's really much too exhausting, you're "beat." And a lively evening follows the experience, an experiment that has just proved to skeptical observers that telepathy really exists, that long-distance communication between two brains is perfectly possible.

The Answer Is in Your Address Book (perhaps)

Well, before we leave the scene, let's have a look at your address book, at the page for last names starting with *N* (like your colleague's). Here's what we see:

Alexander	Ace of hearts	Joseph	Ace of clubs
Andrew	Two of hearts	Joshua	Two of clubs
Anthony	Three of hearts	Juan	Three of clubs
Austin	Four of hearts	Kenneth	Four of clubs
Bart	Five of hearts	Kevin	Five of clubs
Benjamin	Six of hearts	Kyle	Six of clubs
Brandon	Seven of hearts	Mark	Seven of clubs
Brian	Eight of hearts	Matthew	Eight of clubs
Charlie	Nine of hearts	Michael	Nine of clubs
Christopher	Ten of hearts	Nathan	Ten of clubs
Cody	Jack of hearts	Nicholas	Jack of clubs
Daniel	Queen of hearts	Patrick	Queen of clubs
Darryl	King of hearts	Richard	King of clubs
David	Ace of diamonds	Robert	Ace of spades
Dustin	Two of diamonds	Ryan	Two of spades
Eric	Three of diamonds	Samuel	Three of spades
Gerard	Four of diamonds	Sean	Four of spades
Gregory	Five of diamonds	Seth	Five of spades
Jacob	Six of diamonds	Steven	Six of spades
James	Seven of diamonds	Thomas	Seven of spades
Jason	Eight of diamonds	Timothy	Eight of spades
Jeffrey	Nine of diamonds	Travis	Nine of spades
Jeremy	Ten of diamonds	Trevor	Ten of spades
John	Jack of diamonds	Tyler	Jack of spades
Jonathan	Queen of diamonds	Vincent	Queen of spades
Jordan	King of diamonds	William	King of spades
José	Joker		

This sheds a bright light on your telepathic gifts; it shows that the explanation of your mysterious telepathy is amazingly simple. The information about the chosen card is unwittingly transmitted by the person who phones the recipient, whose true name you have given. The last name, anyway; the trick is in the first name. You never state the first name until the card is drawn. It is written down after the card is drawn and after consultation of the address book. The sole purpose of this consultation is to avoid your having to memorize fifty-three first names associated with the fifty-three possible cards (but if your memory is up to it, sure, skip the notebook!).

You may note that there are fifty-three names on the list rather than fifty-two. One of the authors, having been caught by his own brother during one of these long-distance-telepathy demonstrations, thought it prudent to add a name to correspond to the joker (the card chosen on purpose by that wily brother).

After you apparently have verified the phone number, you write the number on a piece of paper, together with the name, including the first name chosen from the list—in this case, Mark. This is the piece of paper you give to the person making the call.

The number of people with all the pieces of information needed to figure out the mystery is reduced virtually to one. The person phoning rarely thinks to tell others that, in addition to information already given, the first name of the colleague has now been provided. It doesn't even occur to them, since they think they have the same information as the others and nothing additional.

In any case, you can casually slip the first name of the person you are contacting into some bland sentence as you accompany the caller for an instant when they step out to use the phone. Then you return to your spot, saying (casually, again), "Okay, it's time for me to see the card you drew so I can concentrate on it." Ten minutes later, what your friends will remember of this series of microevents—if they remember anything—is that you had given the first name, without even being aware of doing so, before seeing the card.

And you can discount another worry. Your friend Mr. Norris has his copy of the list of first names next to his phone and doesn't have to rack his brain to guarantee a high-quality psychic performance. All he has to do is to hesitate a little before giving, slowly, the name of the card.

Possible variants of the scenario have been foreseen. For example:

"Hello, may I speak to Mr. Norris, please?" [Oh no, they didn't give the first name!]

"Whom do you want? There are several brothers here."

"Uh, the one who practices telepathy, we're doing an experiment." [Still no first name.]

"Well, in this family we all do telepathy a little bit."

"I was looking for Mark Norris." [Got it!]

"Speaking."

Nothing could be simpler, as you can see.

We recommend including the real first name of your colleague in your list, in case a lucky break gets you the particular card corresponding to his actual first name. Then you *really* have a complete triumph. You can then say, "Check the phone number yourself in the phone book," which will supply your friends with the first name without any intervention on your part.

Prevention being better than cure, you can also create a specific code for any deck of cards. For example, if you think that someone may ask that you use a tarot deck for the demonstration (which does happen fairly often), you need only extend your grid accordingly, and just keep it all in your address book under your colleague's last initial. After all, his brain really is on the same wavelength as yours, and you've given him a copy of this particular list. And don't forget to put a copy next to your phone because you can be a recipient of telepathic images from your colleague, who will want to prove his psychic powers as much as you do.

Happy mind reading!

Astonishing Feats, Performed Effortlessly and Painlessly

Unlimited Levitation

In Madras, India, in the mid-1930s, the Brahman Subbayah Pullavar demonstrated his psychic powers by levitating in the middle of the street. His helpers held a big cloth in front of him while he got ready. Its removal revealed the Brahman levitating, floating on air with just one hand resting on a cane.

Hindu mystics have no monopoly on the power of levitation. In fact, it was seen in Europe, well before the thirties, in a prettier but perhaps less exotic form. In October 1849, the illusionist Jean-Eugène Robert Houdin presented for the first time his "ether suspension," which quickly became famous.

At that time there was a lot of talk about ether and its applications, and Houdin had the idea of profiting from public infatuation with the subject by augmenting his magic tricks. His description of a public demonstration of levitation comes from *Confidences and Revelations,* the book he published in 1868, which was reprinted as recently as 1980 (Geneva: Slatkine).

"Gentlemen, I tell you, with the solemnity of a professor at the Sorbonne, that I have just discovered a wonderful new property of the ether. When this essence is at its greatest concentration, if you have a living being breathe it in, the body of the patient becomes for a few moments as light as a balloon. . . . "

Discussion being over, I moved on to the experiment. I placed three footstools on a wooden bench. My son stood up on the middle one, and I had him extend his arms, which I supported by means of two canes, one on each of the other stools. I placed an empty flask, carefully uncorked, under his nose, but in the wings of the stage, ether was thrown onto a very hot iron so that the vapor would spread through the room. My son soon fell asleep and his feet became lighter and started to leave the stool. Seeing that it was working, I took away the stool so that the child was supported only by the two canes. This balancing act alone evoked great surprise from the audience. Their surprise grew even more when they saw me remove one of the two canes, and the stool it had been on. The excitement reached its peak when I moved my son to a horizontal position using my little finger and let him sleep that way, extended in the air, and when, defying the law of gravity, I removed the legs of the bench supporting this impossible setup.

Despite the beauty of this suspension by ether, it must be said that the Hindu Brahman can claim priority, ahead of our master magician. In fact, the Brahman Scheschal displayed his powers several years before our "conjurer-physicist-engineer," as Houdin described himself. Scheschal demonstrated levitation on a "bamboo cane and gazelle hide," according to Walter Gibson's 1987 book, *Les Secrets des grands magiciens* (Strasbourg: Éditions du Spectacle). Scheschal used for his sole support something especially soft—amazing, isn't it? Of course, no one ever saw him float up slowly and majestically, nor did anyone see him descend to the ground. And yet this levitation, proving the powers of the omnipotent mind over the gravitation-bound body, did not fail to impress. The diagrams show how a metal rod went up through the center of the cane, along the gazelle hide, and into the magician's sleeve and then was bent around underneath the body to form a seat. For further details we refer the reader to *Magazine pittoresque,* which presented a description and an explanation of how it was done in issue 16 back in 1833.

A Feat of the Heart

If you've just explained the levitation tricks of the previous section to your friends, you're in good shape to go on to bigger things. After all, there's nothing like demystifying a "mysterious phenomenon" to help you substantiate a different one. So you can grab the opportunity to state that the following demonstration can be done only by true initiates who really work seriously on their mind and body and develop their latent psychic powers.

You've met such a person during one of your far-flung voyages. It doesn't matter whether you've really made any such journeys. After all, the renowned Lobsang Rampa, author of *Troisième œil* (The Third Eye)—always in print and a good seller, too—was really named Cyril Henry Hoskins. And he wasn't any more Tibetan than I am, he never set foot in Tibet, and he had

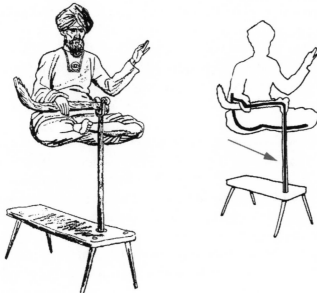

H. Broch after W. Gibson

never even left Great Britain when he published *Troisième œil* in 1958 as an account of his life in Tibet and of his initiation.

Now, this initiate whom you have met—this Wise Man of Tibet—gave you some powers, albeit weak ones, and he taught you to direct your vital energy through pathways in your body. And this is how you are able to control your heart, although your power is still pretty weak, since you are just a simple novice in the first stages of initiation. You can stop your heart by using your mind.

Faced with the astonishment, the smirks, or the evident skepticism of your audience, you explain that it's difficult and exhausting, that it requires total concentration, and that in particular there must not be any noise, but that you would certainly like to *try*—*try*, that is, because you aren't sure to succeed under the present circumstances, which really are not ideal. Then you ask for someone to monitor what happens, someone designated by the group and not chosen by you. After the monitor is chosen, he or she is positioned next to you. Without a word, beginning to concentrate like a yogi, you extend your left fist so that your pulse, and thus your heart rhythm, can be checked. You breathe deeply several times, close your eyes, and fall into your meditative trance. The seconds tick by, silence falls, and then the monitor cries out, "His heart has stopped! His heart isn't beating! It's not beating at all!" And the seconds continue to tick by, more than a minute by the clock. After which, spent and exhausted, you ask to lie down to recuperate. Spent and exhausted but delighted to read on the audience's faces the astonishment, bewilderment, and admiration provoked by your psychic exploit.

A deluge of questions washes over you. You elaborate on your distant voyage, your meeting with the Wise Man, your initiation in the underground sanctuary lit by torches or oil lamps, with sticks of incense slowly smoldering as well. You can say anything, as long as you don't open your jacket or lift your sweater to show the little ball of rubber pinned on your shirt at your left armpit, as long as you don't explain that squeezing your arm forcefully against the ball constricts your artery, thereby reducing your "heartbeat" to zero—a skill not necessarily requiring enormous training or superhuman psychic gifts.

The whole trick to this is getting people to accept the heartbeat and the pulse taken in the wrist as equivalent. This trick is termed the *doormat effect*—the use of a word to designate something other than what the word actually means. Are you surprised by the odd name for this effect? You won't forget it after you think about our observation that doormats often read, "Please wipe your feet," yet you have never taken off your shoes and socks and wiped your feet on one! Q.E.D.

Memories of Youth

Some years ago, one of the authors received a letter that brought to our attention some amusing facts underlying another demonstration of magical powers. The letter, addressed to Henri Broch, was sent from Mexico on March 2, 1992, by R. H. Remy. It recounts an incident that occurred in Mr. Remy's youth, around 1927, which allowed him to become an amateur "fakir."

> In those days, I lived at Lagny-sur-Marne, where my father had one of the big grocery stores. One day a "fakir" came to give a demonstration, lying for a week on a bed of glass shards in a sealed safe. His aides came to the grocery store to get glass shards, and I went with them to our rubbish pile of broken bottles. First they filled two bags, and then they filled a third halfway. The assistants took this bag by the corners and shook it vigorously. When I asked why they did this, they said, "The boss isn't so crazy that he'd pass days on broken glass that hadn't been blunted." As to the safe being sealed, sure it was, but not on all sides, one of which had two sliding panels so that, according to the assistants, "The boss can go to eat and sleep on a bed like everybody else."

Lying on a bed of nails doesn't present any more difficulty for a fakir (or for anybody else) than stretching out on a plank with nails. Students at a zetetics course at the University of Nice–Sophia Antipolis, called "Paranormal Phenomena and the Scientific Method," have already demonstrated such abilities several times. (Zetetics is the application of the skeptical scientific approach to the examination of supposedly paranormal phenomena; for further information you can consult the website at www.unice.fr/zetetique.) Although it takes courage to address such a subject and carry out a demonstration in a university, the risk is nil, or nearly so.

Provided, that is, that the plank has enough nails. That's the whole secret of the bed of nails: the more nails it has, the more comfortable it is. No one, not even the greatest of Indian fakirs, could stretch out on two or three nails. There's no problem, however, when the density of the nails is great enough because individually they exert such weak pressure that they don't pierce the body.

It's So Painful

If you think that the bed of nails may be a bit trite, why not *really* amaze your friends? Take a long, steel needle or skewer, open your mouth, stick out your tongue, and do what one of us did:

© H. Broch

Are you sure you want to try this? Force the long steel skewer through your tongue and stoically show the result to your pals? After all, no one's forcing you; upon reflection you might prefer to continue reading peacefully. But a small bodily sacrifice is necessary for one who wishes to demonstrate his or her mystical powers. The thrust must be quick and forceful, like the scream you give when you find yourself in the painful situation shown in the next photo:

© H. Broch

If you think you can stand it, and your friends don't avert their eyes from your first attempt, then we invite you *not* to try such an experience but instead to use a skewer with the shape you see in the following photo. The whole trick is in the special shape of the skewer, which has a U-shaped part.

© H. Broch

If you like, you can start your demonstration with a normal, completely straight skewer, so that you can show it to the group. However, you will have to exchange it by a trick of your choice, perhaps by accidentally dropping it and picking up the U-shaped skewer instead. You could clean and disinfect a skewer with cotton soaked with bleach, and while you're at it also clean the rest of your set of apparently identical skewers for later use. Obviously, the one that you take out and clean just before starting your show is the U-shaped one. The exchange requires a bit of practice, as well as the creation of a distraction; you need to draw the attention of the audience away from where the key action is really taking place. Alternatively, you could start right out with the U-shaped skewer by holding it between thumb and index finger at the bent part, which would thus be camouflaged.

Practice for a while in front of a mirror and you will see that it's not that difficult to slip your tongue into the U-shaped part while hiding that part. The gasps of horror from the audience will tell you loud and clear that your trick was successful.

If you want to be a bit more impressive or a bit bloodier—but not too much, don't overdo it—you can demonstrate your inability to feel pain by slashing your arms with a knife. Nothing could be easier! Here's the secret method that has been suggested by many people over many years: prepare a saturated solution of potassium thiocyanate and another of ferric chloride. Coat your arm with the ferric chloride and attack it with a knife moistened with the potassium thiocyanate. Big streaks of "blood" will appear where the knife strikes. Once you've wiped off the blood, you can even demonstrate your psychic powers of cellular regeneration by causing all traces of any wound to vanish.

A Candle in the Dark

The "pierced tongue" trick that we've just seen is a classic of the genre and has appeared in books for centuries, as shown in the following picture:

H. Broch

Here you see the means of piercing the head, the body, the arms. A description and a diagram of this classic device appear in a book on witches and their so-called supernatural abilities that dates to the sixteenth century. *The Discoverie of Witchcraft,* dating from 1584, is the work of a person obscure today but deserving of our recognition and that of posterity: Reginald Scot. In fact, the book was republished by Dover Publications of New York in 1972.

The exact spelling of his name is uncertain because the copper plaque in the Church of the Holy Virgin Mary in Brabourne, England, where he is supposedly buried, has the name as "Scott," with two *t*'s, a spelling used by his family for generations. Apparently, Scot used the spelling "Scott" interchangeably on legal documents. He was born in 1538 or earlier and died on October 9, 1599. Lewis Jones published a recounting of "The Discoveries of Reginald Scott" in the March 2000 issue of *Skeptical Briefs* (vol. 10, no. 1, p. 13).

Reginald Scot (or Scott, or maybe Scotte) truly was "a candle in the dark," as he was termed by Thomas Ady in a book by that title in 1656. Scot's *Discoverie of Witchcraft* became famous in the field of magic and illusion because of the twenty pages dealing with magic tricks. The book was singled out—it was condemned as anathema, and King James I of England ordered its destruction.

The Discoverie of Witchcraft is a particularly courageous work because it is a rationalist treatment with the aim of demystifying the superstitions of its day, witchcraft in particular. To understand the ferocity of efforts to destroy the book, perhaps it is important to know that King James VI of Scotland—the future James I of England—was himself the author of a superstitious treatise on demonology. With a genuinely scientific method of research and analysis and with skepticism as a starting point, Scot wanted to open the eyes of his contemporaries, especially judges. It was judges who decided the sorry fate of accused witches and warlocks. Book 13 of Scot's work is a unique source for the history of illusionism, since he reports on numerous techniques for doing the witches' tricks. Many of these tricks can still be learned today by studying the figures.

The shell game is shown by Scot: a coin or a little ball moves mysteriously from underneath one little cup to the other. Tricks with money are also detailed, such as how to make coins appear and disappear, how to make two-headed coins, and the like. We are told how to pass a rope through the nose, mouth, or hand; how to remove a pearl from the center of a string of pearls without untying the two ends; even how to plunge a knife through the arms or tongue or use it to sever your nose. In the next drawing the knife on the left is for the tongue or, with a bigger U-shaped opening, for the arm. The one on the right is for the nose, and the one in the middle is a regular knife, which is shown at the beginning and the end of any demonstration.

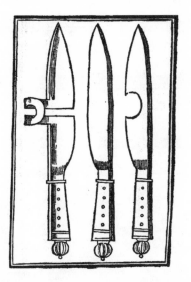

How about driving an awl into your head or through your tongue? Scot tells us what's needed for this, too, as you can see in the following illustration. The leftmost awl is for the tongue, the middle one has a blade that retracts into the handle, and the one on the right is for showing around and for piercing leather at the end of the demonstration to display clearly its unquestionable effectiveness.

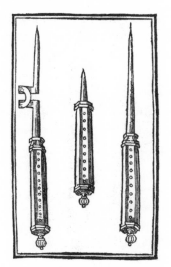

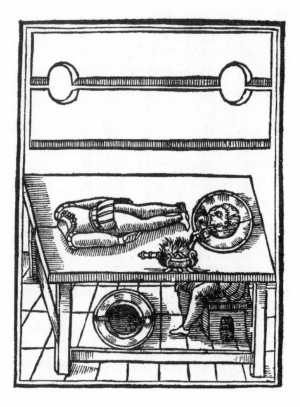

If driving a skewer into your tongue or slashing your arm with big knife strokes seems too tame for you, why not make it more extreme and put on a "decapitation"? Scot offers us a way: the "John the Baptist method," whereby one cuts off someone's head and puts it on a platter while the head continues to live and even to speak. Since a picture is worth a thousand words, the original illustration will serve to explain. For a further re-counting of the talking severed head, see Jean-Eugène Robert Houdin's *Magie et physique amusant* (2002), at www.book-e-book.com.

Walking on Fire

Now that you have become an initiate, we can't resist the pleasure of telling you about a famous and extraordinary phenomenon that requires special powers at least as great as those you have just acquired: walking on fire. This type of walk, on red-hot charcoal, has been observed in many countries around the world since ancient times. Recently, seminars in the United States and Europe have offered the opportunity to "master one's vital energy" and its "alchemical fire" in just one weekend. The seminar reaches its climax with a walk on red-hot coals. And you are told that "a powerful mind can control human tissue so that it doesn't burn when exposed to heat." The real explanation involves several key factors, summarized in the photo below: time, insulation, spheroidal state, heat capacity, and thermal conductivity.

Henri Broch walking on hot coals (Group des laboratoires de Marseille of the Centre national de recherche scientifique [GLM-CNRS Marseille]), May 1992

time

spheroidal state

insulation

heat capacity

thermal conductivity

and the power of the book!

Photo by Yves Bosson

Time

Contrary to preconceived notions, when one walks normally the feet are in contact with the ground for very short periods, less than a half-second per step. Of course, when walking on fire, you don't dawdle on the coals; this is not the time to strike a pose or present one's good side to the camera. On the other hand, you don't have to run, either. In fact, you must avoid running because, when running, you automatically rise on tiptoe, thus placing a smaller surface area in contact with the coals for a given mass or body size. This is a particular problem because that surface is more sensitive. The skin on the toes is generally more delicate and less callused than the skin on the rest of the bottom surface of the foot.

Insulation

The insulation provided by heavily callused skin on the bottoms of the feet is certainly useful, although not essential. However, if you wish to try the experiment, walk barefoot for several weeks beforehand to benefit from this padding.

Spheroidal State

Water in a spheroidal state attenuates heat and is another factor in the ability to walk on fire, although it plays a minor role. What is meant by this phenomenon? A simple experiment you can do in the kitchen will show you. Heat a hot plate just a little, and then throw a thimbleful of water on it. The water will *spread out* and evaporate rapidly, in less than a second. In contrast, if you turn the heat up as high as it will go, wait until it becomes red hot, and then throw the same amount of water on it, the water *won't* evaporate right away. You can use a stopwatch, and you'll be surprised to find that the time for that blob to boil away is more than a minute and a half!

The reason for this is that the water is *not* in direct contact with the plate if the latter is very hot. A cushion of water vapor under the drop is created the moment you throw it down, and this layer protects the drop because the vapor has poor thermal conductivity. Vapor does not transmit the heat of the plate to the water very well.

Some fire walkers moisten the feet before putting themselves to the painful test to gain protection from the spheroidal state of the water. While people can do what they want with their own feet, it seems better, on the contrary, to dry them before the walk to avoid a coal sticking to the foot, which would obviously be very painful.

The spheroidal state of water isn't all that important in walking on fire. But it is in other "fire miracles." It's the main protective factor if you insert your hand carefully (but nervously nonetheless) in molten lead. It's your principal protection if you lick a white-hot knife blade. It's important to emphasize that, in the case of glowing or molten metal, the problem is heat propagated by radiation; that's why spheroidal state plays an important role in such situations. On the other hand, with burning coals the heat is given off by chemical reactions (oxidation), and spheroidal state is less important at the time of contact than is minimization of the oxygen supply to the coals, as we shall see.

Heat Capacity and Thermal Conductivity

Two main scientific principles make walking on fire possible: first, the very low heat capacity of the materials walked upon and, second (and especially!), their poor thermal conductivity (although our feet themselves have a pretty high heat capacity). Together low heat capacity and poor thermal conductivity make fire walking possible; when someone walks on hot coals, the coals are momentarily deprived of oxygen at the point of contact with the foot and momentarily stop glowing at that point.

Heat capacity is the potential of an object to *store* energy in the form of heat, to a greater or lesser extent. Thermal conductivity is the ability of objects to *transmit* heat energy, to a greater or lesser extent.

Let's use the example of a chicken cooking in an oven whose thermostat you have set at 400° Fahrenheit. At any given moment, obviously, the temperature of the interior of the oven is 400°. When you open the oven, however, you fearlessly put your hand in the air inside the oven, despite the temperature of the air there. But you are careful not to touch the chicken, which is at the same temperature. And you are especially careful not to touch the pan in which the chicken is cooking. Everyone knows from instinct—and especially from experience!—that different materials at the same temperature have differing potentials to burn and to transfer heat. The chicken will burn you more slowly than will the metal pan in which you cook it.

If these physical explanations don't work for you, there's always the "power of the book," which may. Henri Broch walked on hot coals, as shown in the photo on page 37, while reading a physical explanation of how to survive walking on fire (from his book *Au Cœur de l'extra-ordinaire* [Bordeaux: Horizon Chimérique, 1991]).

In short, anyone can walk on hot coals with little risk of getting burned. But watch out! "Little risk" of getting burned is not the same as "risk of getting a little burned." It's the risk that is low, but if you get burned, you will get burned *badly!*

Thus, there is no need to appeal to paranormal, parapsychological, or supernatural forces to explain the phenomenon of walking on fire. After all, the authors are familiar with the case of Antoine Bagady, a karate teacher, who walked sixty meters on coals hotter than 1,500° in 1989 without burning himself—and without claiming any paranormal powers. Various people in different eras have tried and succeeded in other unusual experiments: touching red-hot iron with the fingers or tongue; running barefoot on red-hot iron; sticking fingers in molten lead, brass, or iron; and even washing the hands in the molten metal at the opening of a furnace. Examples from Mayne Reid Coe and Jean-Eugène Robert Houdin in particular are described in *Au Cœur de l'extra-ordinaire* .

Of course, it happens that some of those who walk on fire burn themselves. The mental or psychological powers of the burned person have nothing to do with it; it's simply due to the fact that all the factors involved are difficult to control. Moreover, knowing that physics gives you a rational approach to walking on fire doesn't prevent you from feeling pretty darn nervous before your walk. Indeed, knowing the *principles* involved in an effect does not give you knowledge of the *procedures* that must be followed to minimize the risks. And, to top it all off, the encouraging comments of your friends are not exactly calculated to get rid of the butterflies in your stomach: "If you mess this up, don't count on me to push your wheelchair."

We'll feel more comfortable taking a different, efficacious route to this hot demonstration of the claims of gurus and their associates. Let's ask something of those self-styled initiates who supposedly have paranormal powers allowing them to overcome heat and fire. We'll ask them to go bare-foot on a flat, stable surface, which is at the same temperature as the hot coals that they're so proud of walking on. This surface will be an ordinary sheet of copper. We'll ask them to pause (so they don't get tired from walking) for a little while (say, the length of time they allotted for their walk on fire). They may be initiates, but they're not crazy—they will certainly refuse!

Bending Metal

Imagine that you are teaching and you enter an auditorium with about two hundred seats. Facing your audience of students, you start your class. It's a little unusual: a class in "The Powers of the Mind." After forceful declarations and introductory statements about psychokinetic powers that have been demonstrated by different mediums, you launch into a diatribe about these "exotic" powers that are oddly the domain of a select few insiders. And why don't we have similar powers? Why should we think that they are the exclusive privilege of the "elect," of "guides," of "mediums" who, as the name suggests, are really nothing more than "means" in the service of superior beings? Does one really have to have special neurons to bend keys and teaspoons?

"To bend metal using only the mind is a skill anyone can possess, with a little training. Why don't we try an experiment here, together," you suggest. You take a metal wire in your hand and show it around. "We'll try with the end of this iron wire. With your cooperation, we'll all concentrate on it until it unfurls, after having first bent it in various directions to give it any shape we like." Then you bend it. You invite the students in the first row to twist it while you hold the iron wire dangling from your own hand.

"Can everybody see this? Can we start the experiment? You all can see? Yes? No? You can't? Oh, dear." Obviously not many people other than those seated closest to you could clearly see the little section at the end of a wire a millimeter or two in diameter and a foot long. So, you casually suggest: "I have an idea. Let's put it on the overhead projector and then everyone can see it, even at the back of the auditorium. And everyone will see what's going on, nice and big on the screen."

You place the bent wire on the platform of the overhead projector. "Now let's concentrate. And we shall see whether our combined brain waves can do anything or not. Think intently about the wire! Think intently about the wire!"

Nothing much happens and the metal stays unchanged as the seconds and then a few minutes pass by. "Well, maybe we don't have enough power. Let's concentrate even more! One last try!" And then, to the great surprise of everybody present, right in front of their eyes is a psychokinetic phenomenon. The metal wire quivers slightly, then more markedly; it twists around and, gradually, all by itself, takes the form of the letter Z.

What your students have just seen owes nothing to their great mental concentration or yours. It is simply the application of the "shape memory" of a metal that is particularly useful for this type of demonstration. It is called *Nitinol*, an alloy so named because it is made of *ni*ckel and *ti*tanium and was developed by the *N*aval *O*rdnance *L*aboratory of the U.S. military.

"Shape memory" works this way: You bend an alloy with shape memory into the desired shape, and keep it in that shape while heating it. After you cool it quickly, the metal can be bent into more or less any shape while retaining a "memory" of the original shape it was given and to which it will return when it is heated to a certain critical temperature. In the situation we just described, the heat needed for this transition was provided by the projector bulb.

Manufacturers of alloys with shape memory can regulate the critical temperature over a wide range, permitting many uses for these metals. In the December 1994 issue of *Industries et techniques* (no. 755, pp. 70–73), V. Borde published an article called "Alloys with Shape Memory" and enumerated some of their applications:

- devices to mark ruptures in the cold chain needed for safe shipment of foodstuffs
- the creation of a watertight joint between pipes by putting the pipes to be joined inside a sleeve of the special alloy
- fire protection valves, fire detectors, and thermal safety devices for heating systems
- various types of shock absorbers making use of the superelasticity of these materials, including some used in nuclear power plants

- superelastic dentures and dental bridges that would be easier to fit, without the need to tighten them regularly
- surgical staples to hold in place, for example, two pieces of a broken bone using just the heat of the human body: open when cool, the staples close at body temperature.

Alloys with shape memory are even used to protect Italy's Basilica of Saint Francis of Assisi, which was damaged by an earthquake in 1997. Indeed, the metals are particularly well suited to offer protection from earthquake damage. They attenuate and absorb shocks because of their great flexibility and at the same time they support the buildings thanks to their solidity. Structures protected by these alloys can withstand shocks much greater than those that would destroy similar buildings reinforced with steel rods.

Of course, magicians and illusionists keep up with the times, and the first magic tricks using metals with memory date back more than thirty years. And when the metal you've got has a critical temperature of about 80°, the opportunities to demonstrate your powers as a medium with psychokinetic ability really become numerous. Sure, the demo with the overhead projector makes the point, but perhaps this is the best demonstration of all: you bend the "ordinary wire" that you were just using and put it in your hand, with the palm mostly open and facing upward. You put your other hand, palm down, above the wire. Without touching the wire at all, using just your "psychic power," the metal will bend and take a given shape, like a Z, a heart, or a spring (although in reality the heat of your hands will be doing a lot more than your personal magnetism).

But watch out! What you've just read does not necessarily mean that when you see a display of psychic powers, like keys or spoons being bent in Uri Geller's style, these alloys are involved. The methods used by some mediums may be different, and among some mediums we may encounter the evil powers of talented illusionists who have decided to dedicate themselves to dishonest trickery.

There really are specialized sleight-of-hand artists. One of the authors (G.C.) had the opportunity to admire the skill of such an artist, who worked in a fashionable restaurant on the bank of the Seine and stole the watch right off his wrist. Numerous customers were thus relieved of their watches and briefcases, which were returned to them after five minutes.

A medium can also borrow keys or spoons from spectators, bending them surreptitiously without having to employ the subterfuge of the special alloys we've described. One of us (H.B.) did this trick with his students: he asked them to lend him their keys and distracted them long enough for their keys to be returned bent. The method of bending was discreet, simple, effective, and quick: from the batch of borrowed keys, one that had a big opening was chosen. All keys have a flat part allowing you to hold and turn the key; these flat parts have an opening so that the key can go on a key chain. The opening can be used as a powerful crowbar to bend another key by simply slipping the end of that other key into the opening. The students are amazed by this because they have the impression that the illusionist merely brushed against their keys. To those who have keys too strong to bend, there's always the comeback that the key's owner has mental forces too powerful to counteract; this is a tactical retreat for the medium but an ego boost for the subject. It's in just such sleight of hand that the mastery of the prestidigitator is displayed. And there's nothing here to prevent him or her from being honest about it!

Frederick Joliot, a French physicist, was the son-in-law of Madame Curie and won the Nobel Prize in Chemistry in 1935. One day he heard a group of young physicists burst out laughing during a meal. He inquired as to the source of their amusement and was told that one of them had been able to "will" the others to choose an item of a certain color from his hand. Three successes in a row in front of professional rational skeptics bore witness to the reality of this mental power. Joliot smiled and said, "He's cheating. Tell me how it's done."

It matters little that you will always be in a position to point out a fraud. What matters is your maturity and experience, gauged by your ability to resist explanations of unusual, even improbable phenomena automatically, without recourse to your critical faculties supported by acquired scientific knowledge, yet without need of the innumerable fairy tales that humanity has told for ages.

Chapter 2

Amazing Coincidences

Freaks of Nature

A unicorn answering to the name of Lancelot can be seen walking peacefully in a California redwood forest. Amazing, isn't it? But if we look more carefully, this unicorn is far from the mythic image passed down from generation to generation; it is really only a baby goat, a kid. But this is not just any kid—it's the famous inhabitant of an American park, a goat whose forehead tissue ordinarily would have given rise to horns but did not divide normally. This kid has a single, big, central horn.

A few years ago a popular science magazine called our attention to a turtle living happily with two heads and to a snake similarly endowed. And speaking of natural oddities, imagine a lizard six and a half feet long that can cross a lake several hundred yards across, walking at seven miles per hour on the surface of the water! This ability is due to several physical phenomena, including the oarlike effect of the lizard's hind feet accentuated by their rapid movement. The drag of the water is minimized thanks to air pockets created by the feet as they rotate quickly, and when they slap the water an upward propulsive force also helps keep the animal elevated.

Nature's bag of tricks is an inexhaustible source of such surprises. Nature's mistakes, like two-headed turtles and snakes, are no less natural and, if abnormal, are certainly not paranormal or supernatural.

The animal kingdom is not the only source of marvels. Minerals provide mysterious rectangular crystals, with sharp edges several inches long and perfectly perpendicular to each other. Clearly, these must be machine-tooled objects, like something out of *2001: A Space Odyssey*. By what amazing coincidence do we find them in deep archaeological and geological strata? Were they put there by aliens to mark their trip to Earth in a long-ago past? Such fanciful explanations aren't needed; it will suffice to know that this apparently machined block couldn't be more natural. It is pyrite, FeS_2, a compound of sulfur and iron that forms rectangular crystals. The large crystalline form we've mentioned comes from Logroño, in the Spanish Pyrenees.

The animal, plant, and mineral kingdoms offer us plenty of surprises, and there are surely more extraordinary things left to discover in nature itself. What we find in nature is more amazing than the "amazing coincidences" presented to us as supernatural events by those who believe that such coincidences are a phenomenon surpassing nature itself.

"Amazing coincidences"—what sloppy thinking! We would like to show how, with a little reflection and some really simple calculations, you can understand coincidences and realize that often they are not amazing at all. In other words, we would like to show that "extraordinary" things—even the "extraordinary" in our everyday lives—may very well be produced by a cause that could not be more natural.

Psychokinesis? Your Turn to Try!

A Brilliant Illuminating Spirit

The host turns toward the main camera and with a serious, coaxing air looks the viewer straight in the eye and says, "Go ahead! Turn on five or six lights around you!" Then he turns to the medium and asks, "Do you really think you can do it?" After hesitating a few moments, the medium replies, "I hope I have enough concentration this evening, but I'm not doing this under really ideal conditions. To produce long-distance phenomena like this, I usually spend a few days in complete and utter solitude, after rigorous fasting." If he fails, the public will blame the circumstances and not his abilities.

As it happens, the medium does not fail. Light bulbs do blow out in the homes of the viewers of his program, and the viewers call the switchboard of the TV network broadcasting this extraordinary moment "live." As intended, the medium has successfully focused his spiritual power on the material world and blown out light bulbs far away.

Amazing, right? Not so fast! Let's examine this a little more closely. Suppose a million people were watching the show and that it lasts an hour or more. This implies five or six million light bulbs being on for an hour or more. Even if some viewers didn't want to play along or decided to put out some bulbs after a while in the interest of economy, let's say that still gives us two million bulbs that are on for an hour. On average, an incandescent bulb lasts 1,000 hours. This means that, during the length of the show, about 2,000 bulbs will blow out. After all, among the light bulbs installed at random by viewers, there is no reason to think that they tend to be very old or very new light bulbs. Overall, it's as if—among the two million bulbs—there are 2,000 with one hour of life used, 2,000 with two hours of life used, 2,000 with three hours of life used, and so forth, all the way up to 2,000 with 999 hours of life used, and 2,000 with 1,000 hours of life used. Thus, during a one-hour show, those last bulbs will reach the end of their life span and burn out. Chance alone can explain the many phone calls that come in to attest to the power of the medium.

Your Force Crosses Space

Imagine that you have been introduced on TV or radio as a great medium with astonishing psychic powers, psychokinesis in particular, that permit you (among a thousand other things) to control the outcome of a coin toss from far away. Viewers are invited to test your power. The host asks them to take a coin, toss it in the air ten times, and write down each time whether it comes up heads or tails.

The viewers give it a try in front of the screen. The screen shows a closeup of your face furrowed by the immense mental concentration (supplemented by sighing sounds) needed to transmit your force through space. And more than a million people (wow!) jam the phone lines to confirm that, under the control of your thoughts, their coins tended to fall on one side.

The switchboards explode with calls; the host is thrilled; the producer rubs his hands. More than twenty thousand people (that's right, twenty thousand) swear, attest, affirm, and certify that *all* of their tosses, *yes,* without exception, turned up with the same outcome. Ten heads! Ten tails! Of course, it's impossible that you'd get *that* outcome naturally; everyone knows that!

What amazing powers you just displayed there! And yet, your special power really isn't much of anything. Let's look at the situation; let's try some coin tosses. Here is the table that we ourselves get, tossing a coin ten times and repeating that experiment twenty times. It's exactly as if we were twenty different participants.

	1	2	3	4	5	6	7	8	9	10	11	12	13	14	15	16	17	18	19	20
1	h	t	h	h	t	h	t	t	t	h	h	t	h	h	h	h	t	t	h	t
2	t	t	t	t	t	t	t	h	t	t	t	h	t	t	h	t	t	h	t	t
3	h	t	t	t	h	t	t	h	h	t	t	t	h	h	h	h	h	t	t	h
4	t	t	t	h	t	h	t	h	h	t	h	h	t	t	h	t	h	t	t	h
5	h	t	h	t	h	h	t	h	t	h	t	h	h	h	h	t	h	t	t	t
6	h	h	t	h	h	t	h	t	h	h	t	h	t	h	h	t	h	t	h	h
7	t	t	t	h	t	t	t	t	h	h	t	h	h	t	t	h	t	h	t	t
8	t	h	h	t	h	t	t	t	t	h	h	t	h	t	h	h	t	t	t	h
9	t	h	h	h	h	h	t	t	h	h	h	t	h	h	h	h	h	h	h	t
10	h	h	t	h	t	t	t	h	h	h	t	h	h	t	t	t	h	t	h	h
total	5/5	4/6	4/6	6/4	5/5	4/6	1/9	5/5	6/4	7/3	4/6	6/4	7/3	5/5	8/2	5/5	6/4	3/7	4/6	5/5

For each of the twenty people, we show across the bottom of the table the total of heads (h) and tails (t) obtained. One experiment (one person) came up with only one head and nine tails (experiment 7), while another had eight heads and two tails (experiment 15).

Your Turn!

For you to persuade yourself that you can really get such "runs" of heads or tails, we've provided a blank table for you; just put in the total number of heads and tails for each test. The simplest way to proceed is to use ten coins at once, so that you can avoid having to throw one coin ten times, as we did before. The procedure is entirely equivalent. You just take the ten coins in your hand, mix them up, toss them on a table, and count how many heads and tails you get. It'll be just the same as ten successive tosses of one coin made by one person.

	1	2	3	4	5	6	7	8	9	10	11	12	13	14	15	16	17	18	19	20
h																				
t																				
>8																				

STOP!

Really try it! Interrupt your reading, and go get ten coins and a pen!

After filling in the blanks, check off those experiments where you get eight, nine, or ten tosses resulting in the same side showing. The phenomenon that would seem so "improbable," so "extraordinary," will thus be demonstrated before your very eyes; it will actually have happened. In fact, there are very few situations for which this will not be the case. Contrary to what people think beforehand, the chance of getting the same side many times is actually not all that small. The calculation, explained under "Coin Tosses and Probability" in the appendix, shows that we have about an 11% chance of getting at least eight of the same side when there are ten tosses.

Even if we restrict ourselves to the case where all ten tosses give the same result, the probability is still 1 in 512. (Suppose we don't specify in advance whether all ten have to be heads or tails. Then there are two ways to get ten of the same outcomes. We have a $(\frac{1}{2})^{10}$ chance of ten heads in a row, or 1/1024, and the same chance of ten tails in a row, for a total of 2/1024. Another way of looking at it is to say that, with a little more than 100,000 people participating in the experiment, there would be about 200 people with the really extraordinary series of ten heads or ten tails in a row.

The "Small Effect Illusion"

Often we believe that some event has a small probability of occurrence, but we have that impression because of the "small effect illusion." We forget that an event that would be very unlikely in a small number of trials has a much higher likelihood in more numerous trials. It's important to be fully aware of all the trials or attempts, so that rates of success among those trials can be estimated properly.

Using the diagram, we are going to come to an understanding of the essence of research in parapsychology. There are eleven mediums in the bottom row, and they are in the middle of concentrating during some experiment. The subject, methods, and instrumentation employed in this experiment are entirely up to you. Perhaps it involves using cards to predict the future, remote viewing, or telepathy using Zener cards. (A twenty-five-card deck of Zener cards includes five forms—an X, circle, cross, star, or blank space—for testing extrasensory perception.) The type of experiment being conducted doesn't matter a bit; it will have no effect on the idea we are trying to develop here.

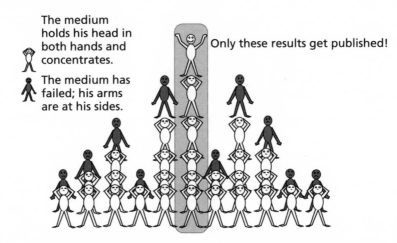

The medium holds his head in both hands and concentrates.

The medium has failed; his arms are at his sides.

Only these results get published!

During the first test, four mediums fail (shaded figures) and seven succeed at obtaining whatever outcome is being sought. During the second test, represented by the third row, two mediums can't get their psychic powers to work and drop out, leaving behind five of their more gifted (or luckier) colleagues. In the third test, there are two mediums not up to the task, leaving three behind in the sample to continue the experiment. Finally, there's a fourth round, and only one of those three succeeds. This individual is acclaimed as "astonishing" and is shown raising his arms in a sign of victory.

Now let's imagine that this medium has happened, by chance alone, to have the desired (and not very improbable) outcome—there was a one in eleven chance of success. Should he really be considered "astonishing"? We need all of the available information in order to decide. In particular, it is important to know that at the start there were a number of mediums undergoing the tests. Often, however, it's just the results of the one successful, gifted person that get publicized.

This simplified example has artificially reduced numbers of mediums and tests, but the point is clear: for a given type of experiment, it is necessary to know the original setup and *all* the results because any conclusion must be based upon the complete set of data and not on a subset restricted in any manner.

The selection of subsets of participants or data is a key issue to consider when analyzing experiments involving so-called parapsychology principles. The context of this sort of experiment and related selection issues make conclusions shaky, although this fact is often ignored or neglected. It's important to realize that the uncertainty associated with data is every bit as important as the data themselves, because the uncertainty determines how reliable the data are and thus how believable the theories based on them must be.

Every aspect of the selection of data must be examined, including the peculiar problem that we are discussing here, namely the nonpublication of negative findings (which is called "publication bias"). Irving Langmuir, who won the Nobel Prize in Chemistry in 1932, illustrates this problem with a story involving his nephew, David Langmuir, and a set of Rhine cards ("Pathological Science," in *CRD, Technical Information Series*, report 68-C-035, April 1968). The cards had been devised by Joseph Banks Rhine, considered by some to be the father of scientific parapsychology. In the 1920s, Rhine (although a botanist) had been the chairman of the Psychology Department at Duke University and had launched studies of telepathy with professional mediums. Later, David Langmuir and his colleagues drew Rhine cards several thousand times and recorded the findings, some rare and "remarkable" but overall displaying expected distributions. They never wrote to Rhine, according to an interview with the Nobelist, and didn't publish anything on the subject.

Of course, there aren't only accidental errors; there's more than selection bias involved in improbable effects that actually occur and get reported. Sometimes there is deliberate forgetting or elimination of data that contradict initial beliefs in parapsychology; we are confronted with data that have been "doctored." So, while researchers on paranormal phenomena often simply avoid clear and concise reports of findings, preferring obscure and windy paraphrases, it's still important to remember that fraud certainly seems to be one of the main sources (if not *the* main source) of research results in parapsychology. Henri Broch has written further on this subject (see, for example, "Struggle for Reason," *Behavioral and Brain Sciences* 10, no. 4 [1987]: 574).

Irving Langmuir observed that Rhine, with whom he had had extensive discussions, had concealed hundreds of thousands of negative results! Langmuir's opinion was straightforward: "No sane person could discard data the way Rhine did. . . . Thus I do not hold his work in high regard."

The illusion of the small effect thus has two sources. One is an error in study design or in the reporting of an experiment, made by an honest experimenter who isn't competent with the data. The other is intellectual dishonesty (cheating), which leads to the deliberate elimination of negative results. From the latter category come the sensational discoveries that for ages have delighted those whose mysterious researches disprove "official science."

It's a Paradox, You Say?

Sometimes—not often, but *sometimes*—there are things that appear normal, usual, and genuine, things that nobody would ever question, that in fact are rather surprising, bizarre, or paradoxical and that arise from the most glaring oddity.

Often—not sometimes, *often*—there are events that seem to arise from the unbelievable, the extraordinary, and the paradoxical; these are events that "shouldn't ordinarily happen." In fact, though, when one examines the situation with a more practiced eye, the events arise from the most ordinary, everyday reality.

A few little forays into the world of Alice in Wonderland will help us to understand this better.

Zoom, zoom, . . . BANG!

It's a common experience: we are vibrating in unison with the deep bass sounds that fill the movie theater while on the screen the spaceship leaves the station and shoots into interstellar space or launches a wild attack on the Black Star, where the Evil Forces are hidden. The room shakes with the force of the roaring motors, and we shake along with it, and we can take part in the glorious spectacle in the starry skies.

But wait a minute! Sound waves can't travel in outer space because there has to be some medium (such as air, for example) to transmit them. Thus, it is paradoxical when the audience is not shocked to hear the noise of the explosions. In other words, you really shouldn't hear anything but just sit and watch the wonderful spaceships and their brilliant, flaming tailpipes move off into the distance—in complete silence.

It's Not Striking

To stick with space examples for a moment, let's think about the news reports from the lunar missions. Who hasn't seen the American flag, planted by the Apollo astronauts, flapping in the wind? But on the moon there's no atmosphere, so there's no movement of air or any gas, so there's no wind and no way that a flag could flutter in the breeze! It's pointless to put forward the crazy hypothesis that you hear sometimes in the United States, suggesting that the moon missions were a big fake and that it was all filmed in Hollywood studios. The movement of the flag on the moon, which everybody could see, was made possible by a wire inserted perpendicular to the flagpole, specifically so that the flag would stay flat and look nice. Otherwise, because of that lack of wind, the flag would always have been hanging down like a rag. In any case, the astronaut banged the flagpole forcefully when he planted it in the lunar soil, transmitting a kinetic impulse to the horizontal wire that was transmitted in turn to the flag. And since there's no atmosphere on the moon, the flag's motion wasn't attenuated by moving through air, and it could continue to wave valiantly for a few more moments.

Conclusion: the flag did flap, but not in the breeze!

Earthrise, Seen from the Moon

The first picture presented in this book is sometimes published with the caption, "Earthrise, seen from the Moon." Although the rising of the blue planet, swirling with water and teeming with life, is beautiful, breathtaking, and profoundly moving when seen from the cratered, desert landscape of its natural satellite, this view is truly paradoxical. In fact, no astronaut has ever seen such a spectacle, and indeed no astronaut ever will, because it is simply impossible. Earth, seen from the Moon, does not rise. The Moon, seen from Earth, rises and sets; it is thus to be expected that the same goes for Earth, as seen from the Moon. At least, that's the underlying reasoning. But it is erroneous reasoning.

In point of fact, the Moon always has the same side facing Earth (more or less), which necessarily implies that when one is on the Moon, at a fixed point, Earth occupies a fixed position. Thus, it follows that there is no earthrise or earthset but rather an immutable view for any moon people who might be looking at us.

The "more or less" is in there because of the phenomenon of *libration*, or little oscillations that the Moon seems to make around its average position as seen from Earth. Because of this phenomenon, over time nearly 60% of the lunar surface can be seen. There are three components: first, libration of latitude, due to the fact that the axis of the Moon's rotation is not parallel to the axis of Earth's rotation. As the Moon moves, the change in the respective positions of the two bodies induces a north-south oscillation. The second component is libration of longitude, due to the fact that the Moon has a slightly elliptical orbit, so that its revolution around Earth occurs at a variable rate (although its rotational speed is constant). The result is a small east-west oscillation. Finally, diurnal libration occurs because the observer on Earth is not at the center of the planet, but at its surface; therefore—because of Earth's rotation—the observer moves from one side to the other of an imaginary line from the center of Earth to the center of the Moon. This, too, results in a little east-west oscillation.

If you really want to see Earth rise and set, you need only go into a low lunar orbit in a spaceship, such as an orbiting command module like those used in the Apollo missions.

An Inside Story

An American scientific team called STURP has been in existence for about a quarter of a century. The acronym refers to the Shroud of Turin Research Project. This group has forty members: thirty-nine believers and a single agnostic. That figures!

On the grounds of the group's composition alone, one might legitimately question its credibility with respect to such matters as the hypothetical shroud of Christ, which has a religious significance obvious to all. (For a detailed study of the holy shroud, on a historical rather than a scientific basis, see the chapter devoted to this subject in Henri Broch's book, *Le Paranormal,* published in Paris by Seuil in 2001.) A simple probability calculation shows that if you were to draw forty scientists at random from the thousands of scientists in the United States, you'd have seven chances in a quadrillion (that is, 7/1,000,000,000,000,000) of getting a group of forty scientists with thirty-nine believers in it! To see how this calculation was made, look under "Probability and the Composition of a Group" in the appendix. And just to clarify and avoid any misinterpretation, a group of forty with thirty-nine militant atheists would arouse the same suspicion.

To put it in perspective, remember that the probability of winning the jackpot in any lottery game is vastly less than that of drawing a group with a membership like STURP's. Thus, in the case of this supposedly scientific organization, probability calculations give us a strong indication that people should take certain assertions of STURP's members with a very big grain of salt.

A Paradoxical Choice

These days, candidates running for public office often rely on the results of surveys to guide their behavior and platforms and to confirm their analyses and expectations. Suppose three candidates (A, B, and C) are on the ballot. Potential voters are surveyed to find out how they would rank the candidates, from most preferred to least preferred. A thousand respondents participate, and the table shows the results.

Order of Preference

first:	A	B	C	A	C	B
second:	B	C	A	C	B	A
third:	C	A	B	B	A	C
Number:	385	370	205	25	10	5

You can see, for example, that 385 people prefer A to B and prefer B to C (first column). Altogether, there are 385 + 205 + 25, or a total of 615 of the 1,000, who prefer A to B. In such a situation, it should be no problem for A to win, if A runs against B. Similarly, 385 + 370 + 5, or 760 of the 1,000 people polled, prefer B to C. No problem here for B if he runs against C—it will be a landslide.

To sum it up, A is preferred to B by 61.5%; B is preferred to C by 76%. So it would seem logical to conclude that A would be preferred to C. But look more closely: A is preferred to C by 385 + 25 + 5 = 415 people, while C is preferred to A by 370 + 205 + 10 = 585 people. So, contrary to what you might think should follow from the first two statements, C is preferred to A by 58.5% of the people surveyed. This little paradox is called the paradox of Condorcet. It can arise in situations where a choice is made among three alternatives, and it is surprising because one expects that a relation expressed in terms of preference would always be transitive. In mathematical notation, this is represented by

$$xRy \text{ and } yRz \rightarrow xRz$$

But a preference relation is not an ordered relation in the mathematical sense.

What, You're a Scorpio, too? That's Amazing!

At a party, you may well find that the person next to you has the same astrological sign—wow, what are the chances of that? But while you are flabbergasted, you know deep inside that since the number of astrological signs is limited to thirteen (and not twelve as some astrologers believe), two people having a sign in common hardly requires a miracle.

On the other hand, suppose that at that same party you are chatting with a delightful acquaintance and it comes up that you have the same birthday. As hardened a skeptic as you may be (or may think yourself to be), it's unlikely that you will completely avoid the thought that this coincidence is a sign of destiny.

The probability that at least two people in any party will have the same month and day of birth is pretty hard to figure out, but at first blush it strikes us that it's got to be really small. That's why, when it happens, we can't stop ourselves from blurting out, "That's amazing!"

Is it really a miraculous coincidence? Suppose we have a group of sixty people. The probability that at least two of them have the same birthday is given by:

$$P = 1 - [365\,!/(365 - N)\,!\;365^N]$$

The calculation actually shows that, in a group of sixty, the probability that there are at least two people with the same birthday is *greater than 99%!* Yes, you've read that right, the chances are 99 out of 100 that it happens. It's the opposite—nobody sharing the same birthday—that would be really surprising.

Do you think 60 is too large a number? Fine. Take a group of 50 instead, and you'll still have a 97% chance of getting at least two born on the same day. For 40, that probability is 89%, and for a get-together of 35, we have an 81% chance of it happening. For a small group, 23 people, we still have a 50% chance. And if you can accept one day on either side of your birthday as "close enough"—still surprising, don't you think?—you could get to a probability that's better than 50-50, with just 14 people! And all these probabilities would be just the same if they were dates of death or dates of anything else you care to name.

As paradoxical (or counterintuitive) as the numbers may appear, they are completely for real. To let you check this for yourself, we present below a calendar that you can use to record the birth dates of the next sixty people you meet. Do this experiment, really do it. You'll be surprised once you've checked all the dates.

Jan.	Feb.	Mar.	Apr.	May	June	July	Aug.	Sept.	Oct.	Nov.	Dec.
1	1	1	1	1	1	1	1	1	1	1	1
2	2	2	2	2	2	2	2	2	2	2	2
3	3	3	3	3	3	3	3	3	3	3	3
4	4	4	4	4	4	4	4	4	4	4	4
5	5	5	5	5	5	5	5	5	5	5	5
6	6	6	6	6	6	6	6	6	6	6	6
7	7	7	7	7	7	7	7	7	7	7	7
8	8	8	8	8	8	8	8	8	8	8	8
9	9	9	9	9	9	9	9	9	9	9	9
10	10	10	10	10	10	10	10	10	10	10	10
11	11	11	11	11	11	11	11	11	11	11	11
12	12	12	12	12	12	12	12	12	12	12	12
13	13	13	13	13	13	13	13	13	13	13	13
14	14	14	14	14	14	14	14	14	14	14	14
15	15	15	15	15	15	15	15	15	15	15	15
16	16	16	16	16	16	16	16	16	16	16	16
17	17	17	17	17	17	17	17	17	17	17	17
18	18	18	18	18	18	18	18	18	18	18	18
19	19	19	19	19	19	19	19	19	19	19	19
20	20	20	20	20	20	20	20	20	20	20	20
21	21	21	21	21	21	21	21	21	21	21	21
22	22	22	22	22	22	22	22	22	22	22	22
23	23	23	23	23	23	23	23	23	23	23	23
24	24	24	24	24	24	24	24	24	24	24	24
25	25	25	25	25	25	25	25	25	25	25	25
26	26	26	26	26	26	26	26	26	26	26	26
27	27	27	27	27	27	27	27	27	27	27	27
28	28	28	28	28	28	28	28	28	28	28	28
29	29	29	29	29	29	29	29	29	29	29	29
30		30	30	30	30	30	30	30	30	30	30
31		31		31		31	31		31		31

Earthquake in Astroland

If someone announced that an astrologer had successfully predicted earthquakes on certain days, some people would be surprised and others comforted by the idea that that person had a proven gift of foretelling the future. Suppose someone told us that the astrologer had predicted that there would be 169 days with earthquakes over a period of three years—1994, 1995, and 1996. On 33 of these dates, earthquakes had actually occurred. We would start to be impressed by this predictive ability.

Is it really reasonable to be impressed? That is, is it justified? If we count only earthquakes with a magnitude greater than or equal to 6.5 or those causing deaths, injuries, or significant material damage, there were 196 throughout the world during those years, according to the U.S. National Earthquake Information Service.

Now we have a little bit more of a solid basis for the following question: Given the 1,096 days in three years, what is the probability that 33 of the 169 dates predicted to have earthquakes would in fact have them, purely by chance coincidence? We'll save you the work of doing the calculation; the probability is approximately 7.1%.

One might feel that this probability is low, lending credence to the astrological predictions. But the astrologer in question didn't announce in advance that he would get 33 successes. If he'd gotten 37, or 41, or 53, or any arbitrary number n, any such outcome would have counted in his favor, so the real question is what is the probability of getting *at least* 33 successes? This probability is the sum of all the probabilities, from the chance of getting 33 to the chance of getting all 169 dates right on the mark. It comes to 30.5%, which isn't small at all!

Again, a picture is worth a thousand words, so we provide the graph on page 60. It answers the question, "What is the probability p that *by chance alone* one would obtain N successes—success being that an earthquake actually took place on the specified date—when predicting 169 earthquakes in a period of three years?" The graph shows every possible case, from 0 days correctly predicted to 169 days correctly predicted.

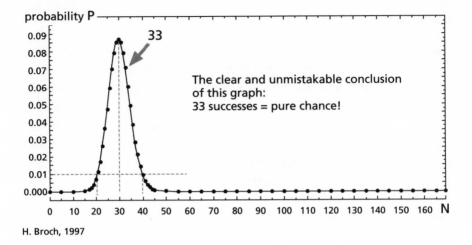

H. Broch, 1997

The graph is noteworthy because it demonstrates that it would have been completely erroneous to consider the outcome as truly extraordinary, beyond expectation, and surely due to the astrologer's work. Indeed, to have *no* "date matches" would have been an astonishing outcome, since its probability is ridiculously small. Similarly, the probability of having only a small number of successes, say fewer than a dozen, is far from a reasonable expectation, as the curve shows. Being rarely right would be evidence for a real power (of "misprediction") on the part of the astrologer in question.

It's also interesting that, if one takes all the outcomes that have a probability greater than or equal to 1%, they would be bounded by a zone running from 21 to 39 successes, and the probability of falling in that zone would be a little above 96% (0.963, to be exact). In other words, you can get yourself a name as a seer or astrologer—simply predict, completely at random, without the slightest psychic power whatsoever, 169 dates for earthquakes over the course of three years, and you will have a little bit better than a 96% chance of getting between 21 and 39 successes. Since there's nothing to force you to remind the public of your false predictions, you have more or less what you need to establish a fine reputation as a serious psychic.

Premonition?

Imagine this situation: You are peacefully lying in your bed. It is 6:04 in the morning, and you are still drowsy. You've hardly awakened when you are struck by the thought of your cousin, whom you haven't seen for years, since he moved abroad, and whom you haven't thought of for a long time, either.

Now it's 6:08. The damn phone rings and you pick it up, only to hear the sad news. Your cousin has died. In such situations no one can avoid making the connection. Here is the long-awaited proof that premonition is for real! Who could deny it now? Such a coincidence is impossible. Maybe dying people send telepathic messages to the living—who knows?

But let's examine the situation a little more closely. We can put the question like this: What is the probability that, having thought about a person, we will somehow learn in the next five minutes, purely by coincidence and without any paranormal influence, that the person has died? To solve that problem concretely, we need to know two things. First, the number of people whose death comes to our attention in the course of a period of, for example, one year. Second, we need to know the number of times one thinks of these people during the same period. Let's do some calculations, very approximate, to lend credence to our results.

First piece of assumed data: you know (in a very general sense, the way any individual knows the president) ten people whose death you learn of over the course of the year.

Second piece of assumed data. You think of each of those people a single time over the same one-year period.

Let's consider one particular person among the ten, whom you have thought about this year. It doesn't matter when you thought of them. Knowing that a year has 105,120 "five-minute intervals," what chance is there that we will be informed of his death during that little interval of five minutes right after thinking about him?

Suppose there's a checkerboard pattern with 105,120 squares, of which only one is red. What are the chances that if I toss a checker at random, with my eyes closed, it will end up on that red square? Of course, the answer is 1 out of 105,120. It is a small probability.

So premonition is a reality, then? Let's not jump to conclusions. First of all, we must also consider the case of the nine other people whose deaths also come to our attention over the course of the year. For each of these, the probability of "having the thought, then learning of their death" is calculated the same way and obviously ends up being the same number, 1 out of 105,120. The chance that you see such an event at all, pertaining to any of the ten people, is given by the sum of the ten probabilities. We conclude that this overall probability is 1 out of 10,512. This is still a tiny chance.

But you have to remember that there's nothing unique about you in this respect; others have brains and can think of their acquaintances, too. Which leads us to say that, taking the U.S. national population and excluding little children, the thought-notification connection must occur each year to about $1/10,512 \times 250,000,000$, or about 23,782 people! So, by chance alone, there are 65 cases like this each day in the United States! This is where these cases come from; this is the grist for the story mill, especially when you consider that we have made starting estimates that are very conservative compared to the actual underlying probabilities. In other words, it is practically impossible to avoid finding, among our acquaintances, at least *someone* with the experience in question.

So this type of "premonition" is actually very prevalent and has no paranormal component to it. If, by chance, it never happened, now *that* would really be a paranormal phenomenon.

Solar Halos, Comets, and Other Signs from Above

Leaving the house one day in May 1995, one of the authors observed a remarkable atmospheric phenomenon in the sky: a big halo around the sun. The sun was encircled by a glowing ring due mainly to the refraction of light by little ice crystals suspended in the atmosphere. After going back in the house to get a camera, we took a few snapshots so we could show this surprising phenomenon to students.

Truly unbelievable, right? Imagine happening to come across such a rare phenomenon when just leaving the house in the ordinary way. This phenomenon has a low probability of occurrence, but was the experience really that extraordinary?

It's a surprising phenomenon, all right, but any other unusual phenomenon would have merited a photo. And such a class of events is innumerable: two intersecting halos in the sky, a cloud shaped like the face of Christ, clouds in the shape of a bike or a car, or the sun appearing between two illusionary suns. This latter phenomenon occurs when light is refracted through atmospheric ice crystals and reflected by them, both at the same time. Then, too, there's rain from a cloudless sky, huge hailstones, hoarfrost at midday, a rain of frogs, a cat and a dog side by side on the roof, a family of geckos perfectly lined up on the windowsill, a bear walking down a suburban street, a tornado all by itself in an otherwise peaceful atmosphere, a ball of fire slowly moving around near the sun, and, of course, we can't leave out solar eclipses. Speaking of which, there's another strange thing that sometimes happens in the skies.

A while back, in southeastern France, there was a truly extraordinary night, filled with huge bursts of heat lightning, so huge that they seemed to portend the end of the world. Many people called the fire department, and more called the local media. The main local newspaper had a big headline the next day: "An Electric Night on the Côte d'Azur."

It was the night of August 10–11, 1999. What? You don't see the significance of the date? That was the night before a famous solar eclipse, which caused a lot of ink to be spilled and a lot of oracular pronouncements among celebrities in Europe. As luck would have it, a specific meteorological phenomenon took place the evening before a notable eclipse, and people felt intrigued by the strange closeness of the two events.

In fact, extraordinary coincidences happen all the time, all over the place; they are very common in everyday life, and everyone has an example to contribute. We don't want to be an exception to that rule, so we'll tell what happened to one of the authors one fine evening in early April 1997.

That evening, Henri Broch had decided to take a few slides of the Hale-Bopp comet, discovered in 1995 by Alan Hale and Thomas Bopp. Visible to the naked eye, this comet had attained its maximal luminosity, so it was pretty easy to photograph without using fancy equipment. And it was important to take the pictures just then, because there would be a wait of 2,700 years for the comet's next appearance.

All set: camera at the ready on a tripod; comet sighted to the northwest over a roof; the shutter is released; the automatic time exposure is activated. The camera itself measures the amount of light striking the film, on an ongoing basis, and nothing else is to be done: just the waiting. The photo shows the result.

H. Broch

The big streak of light going across the picture from left to right, which disappears behind the roof, is obviously not the tail of the comet, which is more to the center of the image. The streak marks the path of an airplane (not in the field of view at the moment the shutter opened), which amazingly just happened to come through in such a way as to "block the path" of the head of the comet.

The probability of this happening is obviously tiny. However, given the number of people around the world who took pictures, perhaps several pictures, of the comet, and given the number of planes that streak across the night sky, maybe the probability is not so infinitesimal after all.

But the more important point is that it wasn't decided in advance to get a picture like this. Lots of other improbable events could have done the trick and amazed us and would have been equally surprising given their improbability. Examples include a single, isolated cloud just blocking the comet; a helicopter with a searchlight going across the field of view (there's a military base nearby); a cat jostling the tripod, resulting in a photo of a cometless sky; a cat climbing on the roof and sitting there, right in front of the comet; a small earthquake (they do occur in the area) making the tripod shake, resulting in an odd picture; fireworks set off by a neighbor celebrating some event (a child's birthday or the occurrence of the comet) and causing a trail of light to encircle the comet; or, finally, the camera breaking down just then.

From the solar halo, to the "interrupted" comet, to the "electric" evening before an eclipse, all of these improbable events have a small chance of occurring, by definition. And that's why, when one of them happens, we are all too quick to pronounce it extraordinary. Extraordinary, yes, but there was nothing to indicate in advance that that particular event was going to happen. And people would similarly call it extraordinary, no matter what other improbable event occurred.

It's important to understand that, when some unusual event or events happen, the chance that an unspecified unusual event will occur is the sum of the probabilities of each of the individual possible events that could occur. And a small probability, plus another, plus another, repeated many times, can give rise to a probability that, if not high, may not be negligible, either.

In other words, when a *specified* unlikely event occurs, that is very unlikely. On the other hand, the occurrence of *some* unlikely event, well, that's likely!

Comparing Risks

As we have just seen, it is intuitively difficult to assess improbable events, those that by necessity have a small chance of occurring individually, but some of which basically must occur because of the large number of events in that class. Likewise, it is difficult to assess the risks encountered daily, weekly, or in any particular period because they are small, even if they add up over time.

The mathematician Sam Saunders at Washington State University clarifies this principle with an analogy involving a frog and cigarettes (reported by K. C. Cole, "Calculated Risks," *Skeptical Inquirer,* Sept.–Oct. 1998, pp. 32–36). If you take a frog and put it in hot water, it will immediately try to escape. But if you put it in water at room temperature, it will stay calmly in the liquid, which you can then heat up without any reaction from the frog. Little by little, you can raise the temperature—*slowly*—by a tremendous amount, and the frog will stay there tranquilly until it gets cooked! In other words, a sharp change in temperature will cause the frog to act, but the same change, if made in small increments over a longer period, will not provoke any reaction.

It's the same with people. We don't mean that you should put your family and neighbors in a big pot of water; we just mean that people often ignore the gradual accumulation of risks in everyday life.

Take cigarettes, for example. Suppose that, contrary to all available data, we could make cigarettes that were completely harmless. Well, this new type of cigarette has absolutely no negative effects on health, except for one little problem, quite rare, but we must mention it all the same. It happens that, because of the manufacturing process, a single pack out of every twenty thousand cigarette packs will contain a single cigarette that is different—different in a bad way. That cigarette is explosive and will kill the user.

Let's look at the bright side. After all, each pack contains 20 cigarettes, and it's only one pack in 20,000 that has a problem. The risk is really low, since someone lighting up has only a 1/400,000 chance of dying from it. The risk is small—but the transition from live to dead is instantaneous. However small the risk, it is the sharp transition that would lead many smokers to quit.

In the United States, more than five million packs of cigarettes are sold each day. This means that, thanks to our new manufacturing process, more than 250 people per day will purchase a cigarette that will kill them that day (if they smoke a pack a day) or within a day or two (if they smoke a bit less). That's more than ninety thousand people a year. "What a slaughter!" you may say. However, it's much less than the annual number of deaths in the United States attributable to smoking-related illnesses.

So it seems that the risks of the exploding cigarette wouldn't be accepted, even though the loss of life is less than that due to regular cigarettes. The unacceptable risk is objectively less than the risk accepted with no problem—smoking regular cigarettes, filtered or not. So you can see that, to be completely irrational, you do not have to dabble in the paranormal.

Looking for Meaning

Look at the picture below. What do you see?

What? You don't see anything? Glance at the following illustration,

with the eyes and mouth enhanced a bit, and then concentrate again on the first picture to "resolve it better" into a recognizable image.

There, that's better, isn't it? You can perceive the face as pretty clearly drawn now, though you may have to choose between Jesus, Karl Marx, a long-haired hippie from the 1960s, or your bearded neighbor. But for the people who saw the "apparition" shown in the photo, there was no need to choose: it was clearly the face of Jesus Christ. In 1985, the little French town of Sierck-les-Bains, near Thionville in the Moselle region, became famous when this simple water stain on the wall of a house attracted the attention of thousands, who saw in it a vision of the Savior; there were even tours that went out of their way to see it. We have the great pleasure of revealing to you, in the photo on the next page, "Jesus of Sierck-les-Bains" as it was, in its original location, in all its glory.

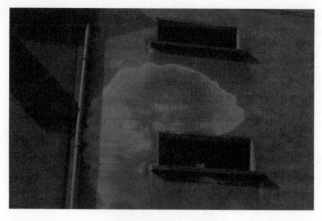

© H. Broch

This calls to mind what we have been saying about improbable phenomena. How many stains, from water or other sources, are there on the walls in a town? In a nation? And why limit ourselves to walls? If, for example, the mark had been observed on the sun (difficult though that would be), the reaction would certainly have been the same.

Among the thousands, indeed, hundreds of thousands of stains, we take note only of those that are recognizable to us, that are evocative of something. It's the equivalent of a Rorschach test, really. Whether the visitation is seen on the scale of a sheet of paper or a wall, our mental process remains the same.

The World Trade Center towers, target of the terrorist attacks whose shock waves reverberated around the world, were continuously filmed for several hours afterward by several cameras. That adds up to millions of images. And, among these, the media could hardly fail to find something. We've been seeing images of Christ; here is the devil.

From the Internet site www.cnn.com (all rights reserved)

From the Internet site www.ap.org (AP, Sipa Press)

In the September/October 1998 issue of *The Skeptical Inquirer* (vol. 22, no. 5, p. 4), a magazine dedicated to the scientific investigation of claims of paranormal phenomena, editor in chief Kendrick Frazier said that one of the greatest attributes of the human brain is a highly developed capacity to recognize patterns in our environment and to ascribe meaning to them. This marvelous ability is useful and productive, most of the time. Human beings seek to understand their environment and ultimately to adapt to it. To that end, this ability to detect patterns is *essential*. It is incontestable evidence of our great intellectual agility. The problem is that we don't know how to control this faculty. Our brain often continues to look for a pattern, for meaning, for sense, even when there is none, and that's when mistakes occur.

Internal, External, or Both?

Finally, we'd like to synthesize what we've been saying: there are basically just two possible sources of unusual coincidences.

The first is *external*. This means that there is a hidden cause, like equipment failure causing measurements or results that are outside normal values or maybe even hoaxes, tricks, or frauds. Although hidden, such causes are often detectable and therefore remediable, even if detecting them would require long and costly investigation or the rejection of certain researchers committed to parapsychology.

The second source of unusual coincidence may be termed *internal*. Such a source may appear at first glance to be easier to control because of the very fact that it's internal. In reality, however, internal causes are difficult to detect, for this requires a self-examination that is not easy. This source of error arises from our deep-seated inability to understand that out-of-the-ordinary events are likely, given a large number of people or cases or a very long time span.

Remember the point we made above. If a *particular* improbable event (that is, one clearly specified in advance) actually happens, that is truly unlikely. On the other hand, that *any* improbable event occurs (no matter which of the many possibilities, limited only by the imagination) is very likely, if not certain. Once again, we must insist: it would be peculiar *not* to see such events.

"What a small world!" "Isn't it amazing that that happened to us?!" "What a coincidence!" Exclamations like these are actually extremely common, and most of the time they are a manifestation of our lack of knowledge about the probabilities of events in everyday life. This lack of knowledge or, more accurately, this misestimation of probabilities is one of the strongest foundations of the belief in paranormal phenomena.

Our view is this: it's clearly evident, from all that we know about our environment and the laws of chance, that extraordinary events (in the literal sense of "out of the ordinary") can occur and sometimes even have a high probability of occurrence. The grand scale of nature itself is enough to elevate those probabilities, and the supernatural isn't necessary.

Chapter 3

Let's Play Detective

Elementary, My Dear Watson

In this chapter, we hope to show you several examples of inquiries one can conduct in the strange realm of the paranormal, the supernatural, irrational behavior, and academic folly. Here are four little investigations in which we can play Sherlock Holmes. But please note that our choice of the famous detective as our symbolic investigator does not at all imply that we subscribe to the beliefs of the originator of that fictional character. Sir Arthur Conan Doyle was in fact a fervent believer in paranormal practices and was much involved as a parapsychologist. He went so far as to write a book, *The Coming of the Fairies,* in which he vouched for the authenticity of photos that two little girls took in a field, showing lovely winged fairies flying in the air. Conan Doyle in real life was a lot less bright than the hero of his detective stories.

First of all, let's investigate the pendulum locator, divining rods, and dowsing. In this investigation, we meet the chemist Chevreul, whose experiments were a model of simplicity and rigor, and then we present a guide for any self-respecting practitioner of dowsing.

If you don't find water using your pendulum locator or divining rod, don't worry, it doesn't matter, since an ossified belief, surrounded for centuries by an intriguing whiff of mystery, can provide water of undoubtedly celestial origin.

Do you fear radioactivity like the plague? Are you afraid of finding a devil in the font of holy water? Have whatever beliefs you like, but do you really think it's okay that decisions on which the future of all humanity depends should be influenced by the alarmist outcries of ideological sects? Is it acceptable that university colleagues—due to laziness, lack of rigor, lack of competence, or love of media attention—should go along with a pack of errors, untruths, nonsense, or lies, and label it "an honorable point of view"?

Well, let's start with a story that will help us to think about this. This story was published by Jacques Poustis in his always-enjoyable column, "Mémoires d'outre-mer" [Memoirs from overseas], in *Science et pseudo-sciences,* August 2000 (no. 243, p. 38). It goes like this: Sherlock Holmes and Dr. Watson are camping. After a solid meal washed down with warm beer and a half-bottle of excellent bourbon, they get into their sleeping bags and fall fast asleep. A few hours later, Holmes awakens and immediately shakes his faithful companion awake. "Watson, look at the sky and tell me what you think." "I see . . . I see millions and millions of stars" is the reply. (We note that Watson must have truly had a remarkable view. The number of stars visible to the naked eye is only in the thousands.)

"What do you deduce from this?"

"*Astronomically* speaking," Watson replies, "knowing that there are millions of galaxies, I observe that there must be billions of planets. *Astrologically,* I observe that Saturn is just entering Leo, from which I deduce— hold on, give me a moment—I deduce that it must be about 3:15 in the morning. *Philosophically,* I conclude that the universe is immense and we count for little. *Meteorologically,* I think that we shall have a glorious day tomorrow. There's nothing else you can conclude in addition, now is there, Holmes?"

Sherlock Holmes, in silence, lights his pipe, takes a long puff, and, looking directly at the self-congratulating Watson, says in an ominous voice, "Clearly, my dear Watson, you really are ineducable, for the first deduction from the view of the sky above is that some hoodlums have stolen our tent."

Pendulum Locator and Divining Rod: Old Hat

It would be exhausting and somewhat demoralizing to repeat this a hundred times in this book and to note yet again that long-disproven phenomena—for which perfectly natural explanations exist—are constantly being presented anew as if they are extraordinary.

And so it is with the divining rod and pendulum, whose boundless merits are constantly being extolled. Why don't those who believe in dowsing test the utility of their methods in the service of a good cause, such as the elimination of land mines, which cause so many horrible, mutilating injuries in adults and (even more sadly) children in countries long afflicted by war?

The Pendulum Swings

Suppose a dowser tells you that his pendulum has swung twelve times, that he has therefore detected a source of water twelve feet underground, and that his method is unbeatable and has never been proven wrong. Rather than undertaking an expensive drilling operation, you may find it more useful to make simple inquiries.

First, ask what would happen if a *French* dowser had looked for water and the pendulum had swung twelve times. Would the water then be twelve *meters* below the surface? Or would the pendulum—by a more remarkable miracle than its mere movement—have made the conversion of units into the metric form? What about a Chinese dowser, coming to visit and practice his art at exactly the same spot, under the same theoretical influence of electromagnetic flows in the earth—would he see his own pendulum make a different number of oscillations yet again, or would the number still be twelve? Would it then be correct to say that water was present at a depth of twelve *li,* or about seven thousand meters? Why should the "electrical fields of the earth" happen to influence dowsing in exact proportion to units of "English measure"? Why would twelve swings of the pendulum indicate twelve feet, or meters, or any other specific human measure?

It's a strange equivalence that the dowser seeks to make between the number of swings, which is a dimensionless measure, and depth, a measure that has a dimension expressed in some arbitrary chosen unit. In Europe, three swings and the water is three meters deep; twelve swings, twelve meters; thirty swings, thirty meters. Aside from the issue of the units, it seems to us that the dowser asserts the existence of an even stranger phenomenon. The deeper the water, the farther it is from the pendulum, yet the more the pendulum swings. Remarkable, isn't it? The force of the action *increases* with distance. If there is water somewhere in the universe, surely the pendulum will swing back and forth an infinite number of times.

These internal inconsistencies (which are nonsense in the literal meaning of the word) don't bother those who believe in the pendulum. They make tremendous claims for their favorite device. The marvels range from the detection of metals in medical diagnostics to the location of oil slicks, missing persons' whereabouts on a map, and many more. To its practitioners, dowsing isn't a science, it's THE science. All you have to do is pose a yes-or-no question, and the pendulum can indicate by its movements whether the answer is yes or no. One is thus in a position to obtain answers to all questions posed by scientists about the world around us, indeed, about anything in the universe.

Attention, scientists throughout the world! Stop your research, stop pondering, stop experimenting, and stop troubling yourselves! Just use dowsing—ask your questions of the pendulum; it knows *everything*.

Sure. However, all the claims of the dowser, including the simple detection of water that we've just been discussing, have long since been debunked.

It Does Swing, Though

We've got to let Michel Eugène Chevreul have his say across the centuries. He is a colleague whose nineteenth-century contributions in this field were definitive. He was born in Angers, France, in 1786 and had had a very long scientific career by the time he died in 1889, in his 103rd year.

In 1812, Chevreul spoke with a famous dowser who used the pendulum as a means of finding things. This led Chevreul to an interest in the possible uses of the electromagnetic fields of the earth. He wanted to reproduce the much-vaunted experiments in this area and undertook several experiments himself with the pendulum. These were not published until 1833. Here is part of his "Letter to Monsieur Ampère on a Certain Class of Movements of Muscle." (It may be found in its entirety in *Revue des deux mondes*, second series, 1833, pp. 258–66.)

Eugène Chevreul
Engraving by Ch. Kreutzberger, in Louis Figuier, *Les Mystères de la science: Autrefois* (Paris: La Librairie illustrée, 1887).

My Dear Friend,

You ask me for a description of the experiments that I made in 1812 to determine whether it is true, as people have told me, that a pendulum made from a heavy object and a flexible string will oscillate when it is held in the hand above certain objects, although the arm holding it isn't moving. You think these experiments are of some importance; in yielding to your arguments for publishing them, allow me to say that it took all my faith in your intellect to get me to decide to put before the public facts of a sort so different from those I have concerned myself with up until now. That being said, I am going to follow your wish to publicize my observations; I will present them in the order that I made them.

The pendulum that I used was a ring of iron hanging from a hemp rope. It was prepared for me by someone with an intense desire that I see for myself the phenomenon that occurred when she held it over water, metal, or a living creature (a phenomenon that she told me about herself). On my honor, it was not without surprise that I saw the phenomenon occur, when, having taken the rope of the pendulum in my right hand, I put it over a vat of mercury, an anvil, various animals, etc. I concluded from my experiments that (as I had been told) only certain objects could influence the oscillations of the pendulum and that placement of others between these influential objects and the moving pendulum would cause the motion of the pendulum to stop. Despite my expectation, I was much astonished when, after having taken a plate of glass, a block of resin, etc., with my left hand and placed each of these objects between the mercury and the pendulum that oscillated above it, I saw the oscillations diminish in amplitude and then stop completely. They started again when the interposed object was removed and ceased completely when the object was interposed again. This sequence of phenomena was repeated many times, with a truly remarkable consistency, and occurred whether the interposed object was held by me or by someone else.

The more extraordinary these effects seemed to me, the more I felt the need to verify that they really had nothing to do with any muscular movement of the arm, although I'd been assured of this in no uncertain terms. This led me to rest my right arm, which held the pendulum, on an adjustable wooden support, which I could move at will to any point between the shoulder and the hand. I soon observed that, when the support was moved from the shoulder to the hand, the movement of the pendulum decreased, and it ceased completely when the fingers holding the pendulum were themselves supported. Moving the support back from the hand toward the shoulder had the opposite effect. As the support passed the same points between the rope and the shoulder for a second time, however, the pendulum's movement was slower at each point than it had been when the support was moving *toward* the hand. After this, I thought it very likely that the phenomenon was caused by a movement of my muscles occurring without my knowledge. I thought I ought to take this idea more into account after I hazily recalled that I had been in a very peculiar state when my eyes were following the motions of the pendulum I was holding.

I repeated my experiment, with my arms completely unsupported, and convinced myself that the recollection I just mentioned wasn't an illusion, for I clearly observed that when my eye followed the swinging pendulum, I had a predisposition to or tendency favoring movement that (although it seemed completely involuntary to me) was more marked when the pendulum was swinging in big arcs. Seeing this, I thought that the results might be completely different if I did the experiment again, blindfolded. That's exactly what happened. With the pendulum swinging over the mercury, someone blindfolded me. The movement soon diminished. Although the oscillations were small, they didn't diminish noticeably in the presence of objects that had apparently stopped them in the first set of experiments. Finally, from the moment when the pendulum stopped completely, I held it for another fifteen minutes over the mercury, and it didn't start moving again. And all during that period, completely unbeknownst to me, someone was repeatedly interposing and taking away the glass plate or the block of resin.

Here is how I interpret these observations:

When I held the pendulum in my hand, a movement in the muscles of my arm, imperceptible to me, moved the pendulum from the resting state. These oscillations, once begun, were soon augmented by the tendency to encourage them induced in me by seeing them. Now, it must be clearly understood that movement of muscle, even if it be due to such a tendency, is nevertheless weak enough to be stopped. I wouldn't say it can be stopped at will. But it can be stopped when someone simply has the idea to try to see whether it can be stopped.

There is thus a close link between certain movements and the act of thinking about them, even though that thought may be entirely different from willful control of the muscles. I think that in this respect the phenomenon I've described may be of some interest in psychology and even in the history of science. They show how easy it is to take illusion for reality, whenever we are concerned with a phenomenon in which our own body is involved under circumstances that have not been analyzed sufficiently.

I had been satisfied to see the pendulum oscillate over certain bodies and had accepted the experiments in which the oscillation was stopped by the interposition of glass, resin, and other objects between the pendulum and the object that seemed to cause its movement. I certainly had no reason to doubt the powers of the divining rod and other objects of that type. Now one can really understand how people of perfectly good faith, otherwise enlightened, are sometimes driven to chimerical ideas to explain observations that don't really arise from the physical world as we know it. [At this point, Chevreul inserts a note that we will discuss after the end of the letter.]

Having become convinced that nothing extraordinary was going on in the events that had so surprised me, I developed a very different point of view from the one I'd had when I first observed those phenomena. So different, in fact, that for a long time afterward, at various points, I tried to get these phenomena to happen again, always without success. In recalling your own report of a phenomenon that I had seen with my own eyes, a dozen years ago, I show our readers that I am not the only person for whom sight influenced a hand-held pendulum. No doubt you remember one time when I was at your house with General P. and some other people, and my experiments became one of the subjects of conversation. The general wanted to know the details, and after I filled him in, he made it clear that the influence of sight on the pendulum was contrary to all his thinking. You remember my suggestion that he do the experiment himself. He was struck by astonishment when, having had his left hand over his eyes for a few minutes and then removing it, he saw the pendulum in his right hand completely still; yet it had been swinging rapidly at the moment he started to block his view.

The preceding facts and my interpretation have led me to associate them with other, everyday phenomena: by such linkage, the analysis of the latter will be made simpler and more accurate than ever before and will permit the assembly of a collection of facts whose overall interpretation lends itself to a general, unifying theory.

What about the Divining Rod?

In his letter, Chevreul inserted a note making the connection between the action of the pendulum and that of the divining rod:

I can well understand that a person of good faith, with attention focused exclusively on the movement of a hand-held divining rod, could for reasons unbeknownst to himself manifest the tendency to move that is necessary to cause the phenomenon that he seeks. For example, if this person is looking for a spring, and if he's not blindfolded, and he sees a verdant patch of grass while he's moving along, an unconscious muscular movement may cause the divining rod to move because of the connection between the thought of lush vegetation and the thought of water.

For Chevreul, the analogy between the two situations was clear, and the explanation of the involuntary effect of thought on movement, which was made for the pendulum's actions, applied for the divining rod as well. The influence of thought and the need for vision in order for the phenomenon to work have been demonstrated many times since Chevreul's day, but the claims of dowsers haven't abated much.

The 200,000 Euro International Challenge offered a prize to anyone able to prove the existence of any paranormal phenomenon whatsoever. Claims were many, but few dowsers wanted to offer a simple demonstration of what is supposed to be their primary claim, namely their ability to detect water, preferring to limit themselves to other types of experiments. Nevertheless, there was one who decided to face such a test.

This competitor for the prize declared that he had readily detected water with his divining rod and that he'd provided such services to many satisfied customers for many years. Moreover, of one hundred wells dug at his direction, ninety-eight provided water. As to the other two, the drilling company didn't do their job correctly: they drilled off the mark in one case and too deeply in the other, breaking into an underground vault into which the water was lost.

The competitor's test took place on a lawn on the campus of the faculty of sciences at the University of Nice on July 12, 2001. The dowser first chose, using his rod, a large area where he said there wasn't any "tributary subspring" that could be distracting—it might interfere with detection. Then the experiment began. A hydrant, at a distance to avoid "interference," of course, fed water into a long pipe. Underneath the chosen experimental

Diagram of apparatus for test of dowsing, Science Faculty of Nice University, July 12, 2001

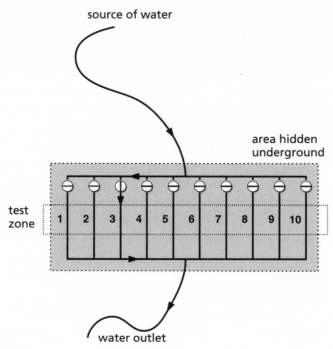

source of water

area hidden underground

test zone

1 2 3 4 5 6 7 8 9 10

water outlet

H. Broch, 2001

area, this pipe was split into ten, each with a shutoff valve, and all ten pipes fed into a common outflow pipe. The diagram illustrates the setup.

The ten pipes were hidden, but their paths were numbered and their locations known precisely. A single valve, chosen at random, was set in the open position. The dowser knew the setup and simply had to indicate, by using his divining rod, which of the pipes had water running through it. He could verify that the water was really circulating inside the apparatus, at the flow rate that he had requested, since the single, common exit pipe was in plain sight where anyone could see it and check its output. The divining rod could practically touch the pipes, which were a dozen centimeters apart. That separation is much more than the one-centimeter separation that the dowser assured us was the minimum necessary for detection in this experimental setup.

There was a one in ten opportunity to pick the right pipe just by chance. The dowser was sure of himself and announced that, under such experimental conditions, he should have a success rate of nearly 100%. No problem.

The experiment was repeated twenty times, with pauses as needed. Twenty times the rod reacted clearly and distinctly, after a few tests necessary for the dowser to check on the sharpness and reliability of the rod's reaction to a piped water source. The experiment was filmed, and the results were duly recorded. What do you think the success rate was?

Two correct choices in twenty attempts demonstrated completely chance findings and the total failure of the claimed capability of the water-detecting rod. The fakery of the lovely, warm, friendly practitioner—whose good faith was never doubted—was clearly evident. It was otherwise completely impossible to understand his performance.

"But I'm telling you, it really does work," he told us, "I can detect water. Wait! Look at the intake pipe where the water flows in. I'm closing my eyes, I'm really closing my eyes . . . "

He closed his eyes, walked forward, and the rod turned in his hands with a sharp movement and pointed to the exact spot on the soil where the intake pipe was located.

"Fine! Want to try again?" he was asked. He closed his eyes and moved around, and the divining rod moved yet again in his hands, pointing to the exact location of the pipe that used to be the one with the hidden water flow. But the diviner did not know that the valve settings had been changed while he had been walking around. And no "persistence of effect" was invoked as an explanation of this gross error. The first impression of the position of the remembered pipe was clearly the determining factor.

Later, on the suggestion of the experimenters, the dowser was put to another test, by a member of his family, with an extremely simple arrangement involving a single pipe. He simply had to determine with the rod whether water was flowing or not. Thus, he would have a 50-50 chance, with the obvious assumption that he couldn't see the pipe vibrating, hear water flowing out of it, and so forth—an extremely simple test, without skeptical scientists around who might emit negative "vibes." One parent to open or shut a far-off tap so that the dowser couldn't see the movement.

These trials then took place; they showed no tendency to detect water using the divining rod.

There you have an example of a person who, completely in good faith, had over the years convinced himself—and others—of his special power, which was never of a magnitude exceeding what would have been expected based on the laws of chance.

Do you see what's happened? The pendulum, the rod, and all the other dowsing equipment have given rise to various suggested explanations, without the existence of the phenomenon having first been proved! Why put the cart before the horse? All the work in this area to date—every scientifically controlled experiment—has shown that you can turn a completely successful attempt by the dowser into a total failure simply by taking away the information about the location that is to be identified. Take away that information, and the results are the same as those anticipated from a knowledge of probability.

Although you're not a dowser now, you might, despite the preceding discussion, decide to change professions and take up dowsing, which (we freely admit) may sometimes provide a better living than does scientific research. If this should be the case, we would suggest that you familiarize yourself with techniques for influencing responses, which all self-respecting mediums, dowsers, and parapsychologists must know. Such techniques can be used to induce a biphasic response effect, caused by changing the rules of the game according to who is playing or their reactions. Although it may seem pretty surprising that some people can do this and others will accept or fail to detect such a trick, it actually happens pretty often.

Here's an example. Do a quick survey of responses to the following question: Would scientific support for paranormal phenomena be a substantial argument in favor of the latter? Without a doubt, you will get nearly unanimous affirmative responses. Pose this additional question to the same people you just finished questioning: If science were to reject the paranormal, would that weaken your belief in the latter? Now people will answer No, with equal unanimity. It's as if you apply for admission to a club—if you are accepted, it's a good club; if you are rejected, it's a lousy club.

This biphasic response effect is often found among believers in pseudoscience and even more so among those who believe in alternative med-

A Little Guide for the Dowser

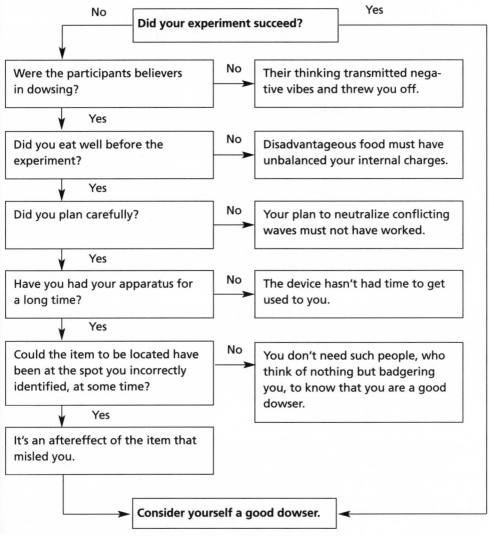

H. Broch and D. Briant, 1999

icine. (There is some overlap among these groups.) To help you practice the special language of the pseudosciences, we'll end our foray into the land of pendulum and dowsing rod with a little guide to fallback positions for every good practitioner. This is selected from among the possibilities examined by Henri Broch and David Briant (the latter a student of skeptical philosophy and physical chemistry in 1998–99, working on dowsing).

The Mystery of the Sarcophagus at Arles-sur-Tech

Arles-sur-Tech, a small town in the Eastern Pyrenees, ought to enjoy worldwide renown: believers and nonbelievers in miracles ought to rush there, not to mention scientists with a special interest in the details—physicists and chemists, geologists, hydrologists, practitioners of the divining rod and the pendulum. The church there possesses the unlikeliest of springs, a water source a thousand times more mysterious than the dowsing rod, which rises from somewhere between heaven and earth. The spring originates in a marble tomb, connected with the ground only by twin pedestals, each eight inches high.

This text, from the first half of the twentieth century (O. Leroy, *Vie intellectuelle* 10, no. 43 [1936]: 191), did not bring renown to this odd tomb. It would be another half-century before fame arrived. But first, a few words about the monument itself.

The tomb—or more precisely, sarcophagus—is carved from a block of marble six feet long, two feet high, and a foot and a half deep. The lid is in the shape of a prism and is one foot from base to peak. The sarcophagus, which some claim dates from the fourth or fifth century and which is supposed to contain the remains of Saints Abdon and Sennen, became famous in Europe after a television program was broadcast concerning this "holy tomb" (as locals have long called it). As shown in the photo, the holy tomb is in the open air, at the base of a wall forty feet high that overlooks a small courtyard. This courtyard opens to the north, facing the adjacent church. The sarcophagus cover is as thick as the wall—about thirty inches—and it does not touch the wall completely. One can even slip fingers into interstices in two or three places. The sarcophagus does not rest directly on the soil but on two blocks of marble.

© Henri Broch

The "miraculous" phenomenon associated with this sarcophagus is as follows: each day, a pretty good amount of water—about a quart—accumulates inside of it. This water is practically pure, and people say that it has curative powers. You can draw it out through a little hole in one of the sides, between the sarcophagus and the lid, a space in which you put a little siphon.

It even happens that the sarcophagus overflows. The water production at times has reached eight hundred quarts per year. There is no apparent plumbing or external water supply to account for this. The television show called it "an unsolved mystery." During this show, various documents were presented, including interviews and a study conducted in the late 1950s by hydrologists, in support of the conclusion that "studies up until now leave something to be desired" and "the holy tomb does not yield its secret."

A plaque attached to the courtyard's iron gate has an inscription providing a thumbnail sketch of the history of the monument. It, too, states that "the holy tomb has not given up its secret."

© H. Broch

An Inquiry and a Blackout

So, is it a mystery? A miracle? Well, it's neither!

We've devoted some pages here to the mystery of this sarcophagus, and complete information may be found on our website (www.unice.fr/zetetique). This extensive treatment is not needed because the mystery is so profound, or because a big investigation is required, but because the situation is pretty typical of the sometimes negative effect a powerful medium such as television can have on several levels. In fact, many people still believe that the mystery of the sarcophagus is unsolved; some even have recourse to explanations that actually aren't explanatory, since the solution was set out in detail more than forty years ago.

In fact, contrary to what was stated explicitly in that TV program and in various publications, the scientific investigations carried out thirty years ago already provided a conclusive explanation. We refer to the results of three hydrologists—Pérard, Honoré, and Leborgne—from whom we have borrowed extensively here. These results were published in the December 1961 issue of the journal *La Houille blanche* [Hydroelectric Power] (no. 6, pp. 873–81), in articles signed by G. Pérard and C. Leborgne.

The investigation was carried out with the agreement and cooperation of the parish priest of Arles-sur-Tech, who gave the researchers the key to the courtyard, and with the collaboration of Monsieur Rougé, a retired schoolteacher. In 1961, for two and a half months, measurements, observations, and experiments were carried out according to a plan established in advance. This was interrupted only for a few days around Easter to allow for visits by the faithful and by tourists.

A list of possible explanations was stated in advance:

1. capillary action involving moisture traveling up into the sarcophagus via its supports
2. condensation of water out of the air during the warmer times of the day, when the temperature of the stone sides of the sarcophagus is cooler than the air around it
3. a phenomenon relating to dew: cooling of the sarcophagus overnight, due to heat radiation, together with lowering of the temperature of neighboring air layers, resulting in the deposit of water droplets
4. the complement of the two preceding hypotheses: condensed water (and maybe rainwater, too) traversing the lid because of capillary action and gravity

The measurements carried out:

1. temperature (recording thermometer placed near the sarcophagus, with the tracing collected every week)
2. humidity (recording hygrometer placed next to the thermometer)

3. the level of the water in the sarcophagus (measured in a graduated vessel in a tube linked by a siphon to the interior of the sarcophagus)
4. wind direction and speed
5. rainfall

The experiments performed on-site (others were done in the laboratory):

1. cementing the cover shut to see whether the water came solely from the air circulating within the sarcophagus
2. the placement of a nylon cover on the sarcophagus lid with a two-inch space for air circulation

Two months without rain coincided with two months without change in the water level in the sarcophagus (except for declines caused by the priest taking samples). This fact alone is revealing. It shows, as the technical report says, "that the sarcophagus may produce between one and two quarts daily, and the production is therefore not absolutely continuous, as had long been known." On April 10, 1961, rainfall was 0.22 inch, the next day 0.27 inch, and the day after that, the level of water in the sarcophagus changed, going up by about 0.039 inch. These measurements and those of the subsequent days through April 23 are given in the graph.

Sarcophagus at Arles-sur-Tech: Water Level as a Function of Date

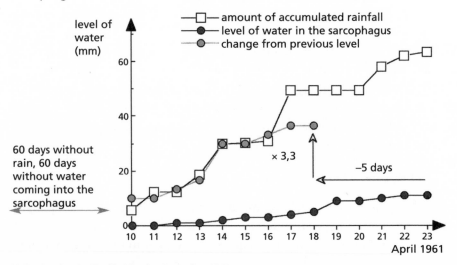

H. Broch, after *La Houille blanche*, December 1961

The lines in the graph indicate the amount of accumulated rainfall and the level of water in the sarcophagus. They show very clearly that the sarcophagus gets refilled by rainwater. The hydrologists "arrived at the conclusion that the water takes on average five days to get through the lid, and on average a third of the rainwater makes its way into the sarcophagus."

A peek into the interior of the sarcophagus, through whatever little cracks allowed it, had in any case already revealed the presence of large water droplets gathering at certain spots on the underside of the cover. Rain had preceded this observation by twenty days, showing that the complete outflow of rainwater from the lid can take pretty long compared to the mean.

Water striking the cover drop by drop disappeared almost immediately, making an expanding circle of moisture on the surface. The surface is very much at an angle, and the damp circle had its center at the point of the droplet's impact. The surface is irregular, with little hemispheric pits each about 0.05 inch in diameter; once filled, the pits empty in about forty-five seconds.

In that section, the hydrologists' study demonstrates that certain descriptions are simply wrong. For example, when it is said or written that "sometimes the sarcophagus overflows," you think that there is at least a tiny stream of water running over the side. The reality is rather different, though, since the phrase comes from a statement signed by ten people on April 3, 1942, which says, "The sarcophagus is full; the liquid overflows, a big drop falling every two minutes on the front of the tomb." The tomb rests at a slight angle, which explains the overflow being restricted to a particular point.

The overall conclusion of the 1961 technical report on the Arles-sur-Tech sarcophagus was as follows:

The cover of the sarcophagus is permeable, and rainwater enters it, takes an average of four to six days to get through the stone, and then passes drop by drop into the interior. As there cannot be any substantial circulation of air between the inside and outside, there is practically no evaporation, and thus the water can accumulate to a significant extent. Moreover, since the rainwater washes and even slightly erodes the sur-

face of the cover, the cover remains clean and permeable, and the phenomenon can go on indefinitely.

. . . Then why does the water remain in the sarcophagus if the main
portion is made of the same marble? First of all, the stone used does not
look exactly the same, and it is possible that it was cut from a different
slab, of low permeability. Also, the still water in the sarcophagus allows
the deposition of the most minuscule particles that it might contain, as
well as the little bit of dust that manages to get through the interstices.

. . . There's also the possibility of a little bit of dust carried along by water
that trickles down on the cover and gets in between the cover and the
main box of the sarcophagus, by the mechanism of hanging droplets. . . .
Over the course of centuries, the deposits must have made the sarcophagus watertight by clogging the very pores of the stone.

The conclusions also indicate that, since the cover is permeable, the contribution of dew may be significant, for water deposited on the cover can
then get into it.

In short, as the investigative team said at the start of their work, "We
have worked, pondered, probed, felt with our hands, siphoned, and we
have put our finger on the drop that fills the sarcophagus."

The "amazing" phenomenon at Arles-sur-Tech, brought to the attention
of the general public by a European television show called *Mysteries,* is
thus, in point of fact, a phenomenon that couldn't be more natural. After
the investigation by the hydrologists in 1961, it wouldn't have attracted
any more attention from anybody had it not been for the pitiful and re-
grettable misinformation conveyed by the TV show's producer and journalistic host. A relevant anecdote will illustrate their obscurantist bias.
Months before the broadcast in question, while it was being produced, a
journalist working for the production company called one of us to ask
whether we'd like to be a scientific consultant for them. The conversation
touched on the sarcophagus at Arles-sur-Tech, among other subjects. The
journalist was given detailed information, in no uncertain terms, explaining the sarcophagus's water supply and that it had been a long time since
it had been a mystery.

We heard nothing from them after that. And later on, during the premier of the *Mysteries* show, the phenomenon of the sarcophagus was presented as an unsolved enigma.

Abundant Nonsense

Another result of the *Mysteries* broadcast was to provide fodder for parapsychologists. Here are some selections from the recent bumper crop of stupidities.

Yves Lignon made these statements in the French newspaper *Midi Libre* on July 27, 1998:

> The solution suggested in several articles that appeared between 1959 and 1961 can easily be refuted for several reasons. According to the authors of those studies, the sarcophagus gets filled by the infiltration of rainwater. Perhaps, but when you read that the measurements of water levels are made using a schoolboy's ruler, you ask yourself, even if not trying to be combative, whether you can really take such people seriously.
>
> . . . If . . . you do . . . some calculations based on the tables of numbers provided, you will realize that there's nothing to support the statement that there exists a statistical association between the amount of rain and the level of water in the sarcophagus.

The journalist who wrote the main part of the article, J. Vilaceque, told us an astonishing fact about the sarcophagus: "It is there under its own covered porch in a courtyard to the left of the church, set on two pedestals of stone twenty centimeters high" and gets filled with water "all by itself. Rain does not touch it."

Yves Lignon came back to this point again in a book, *Les Dossiers scientifiques de l'étrange [Scientific studies of the unknown]*, published in Paris by Michel Lafon in 1999, and also in a radio presentation, *Vent sud [South wind]*, on September 20, 1999. It can't be a matter of rainwater because the sarcophagus "is sheltered," according to the first publication, and also because "studies of the water, in England," have determined that the water isn't rain, according to the second.

How to Fabricate a Mystery

Where did the journalist from *Midi Libre* and where did this guy Yves Lignon observe that the sarcophagus is sheltered? It's absolutely false! The sarcophagus is in a little courtyard, out in the open, although it isn't strictly true that there is nothing above it. There is no covered porch. Quite the contrary, there is a slight bulging of the tiles from the wall behind it (as seen in the photo previously discussed), which serves to direct *even more* water toward the sarcophagus (note that we say "toward," not "on top of").

Either these two people never set foot in Arles-sur-Tech and blithely made up "good copy," or they saw it and deliberately and brazenly lied. Although the question of the credibility of such people can be settled quickly, there are still readers of their work who, in good faith, believing themselves to be objectively informed, remain convinced that the sarcophagus is sheltered from the rain and that its water source is a complete mystery.

That "Schoolboy's Ruler"

It seems that Yves Lignon does not understand the basic point that, with a schoolboy's ruler, one may easily do accurate work. Is the problem that he thinks that the markings on kids' rulers are in "sillimeters"? A millimeter is a millimeter, and a school ruler will more than suffice for this type of measurement.

Some parapsychologists seem unaware that you can do experiments perfectly well with sticky tape and bits of string that yield, as a first approximation, completely sound results with which to explain a phenomenon or test a hypothesis. It's the experimental design and not the equipment used that's important; in point of fact, the *way* in which a measurement is made is more important than the quality of the instrumentation used. It doesn't matter that that measuring system is primitive, as long as it works. And it did!

A Fabulous "Statistician" in Action

Yves Lignon says, "If . . . you do . . . some calculations based on the tables of numbers that are provided, you will realize that there's nothing to support the statement that there exists a statistical association between the amount of rain and the level of water in the sarcophagus." He must be dreaming! The association between the amount of rain and the sarcophagus's water level is, on the contrary, *certain*. It's obvious, if you do the calculations correctly—that is, taking into account all the zeros in the two months before the first changes in the water level, as you can see on the graph. These are values that must be included in the calculation testing the hypothesis of a correlation between rain and the water level. Two months without rain, two months without variation in the water level (other than the priest's sampling, of course)—and then, once it rained, the level started to rise. Hardly a complicated analysis: the link is there.

Reinventing the Wheel

Condensation of water is "truly a promising avenue for research." Or so a parapsychologist informed us about the great discovery he had made in 1998, while reading an article in a local newspaper, in which a physics professor suggested that condensation should be looked at as an interesting alternative explanation for the phenomenon of the sarcophagus. This must be a dream, too.

The 1961 article by the hydrologists mentions condensation of water vapor from the air, the first explanation that everyone involved with studying the phenomenon had originally considered. The very same article states, concerning temperature readings, "Those data were used to do a few simple calculations of the amount of water that can collect due to condensation." It seems that these "simple calculations" are beyond the capacity of certain great psychics who, moreover, do not know that a cooled stone wall can produce only a small amount of water compared to the amount involved in the phenomenon of the sarcophagus. (More about this later.)

To appreciate the "discovery" of condensation a bit better, we present a few lines (out of many) devoted to the great mystery at Arles-sur-Tech. "Others have thought that the stone of the sarcophagus could effectively absorb moisture from the atmosphere, and that the water collected at the bottom is nothing but that moisture, deposited by condensation. But . . . it must be acknowledged that this type of process could never produce hundreds of liters per year." This text, as clear as pure water, dates back to sixty years—that's right, *sixty years*—before the splendid new hypothesis that fills Lignon with so much enthusiasm. (The source of the quote is O. Leroy, "Un Prodige permanent: La Tombe d'Arles-sur-Tech," *Vie intellectuelle* 10, no. 43 [1936]: 191–96.)

To avoid any ambiguity and to demonstrate clearly that the condensation hypothesis is not at all new, we remind you that it was a possibility acknowledged by practically everyone involved seriously with the search for a natural origin for the water in the sarcophagus. Differences of opinion exist only with respect to the relative importance of the condensation in the accumulation of the water.

Some people still favor very-high-yield condensation of moisture from the air and reject the rainwater hypothesis. This is in complete contradiction to the data that have been available since 1961. Others accept that the explanation is primarily the rainwater, as shown by the 1961 publication, with some minor contributions from other sources.

As an example of the many texts dealing with the subject, here is an excerpt from an article published in the *Mémoires de l'Institut de préhistoire et d'archéologie des Alpes maritimes* in 1975–76. The article deals with the air as a water source (C. R. Cheneveau, "L'Eau dans les castellaras de la Ligurie marinalpine [puits aériens]," *Mémoires* 19 [1975–76]: 3–16). The authors of the present work thank Denis Biette for calling the article to their attention. Here is what it has to say about the sarcophagus of Arles-sur-Tech:

[The water in the sarcophagus] is produced partly by condensation of humidity from the air, resulting from the difference in temperature inside and outside the tomb. Perhaps there is also some produced by penetration on rainy days, water making its way through the porous lid, or by little droplets slipping in through such interstices as exist between the lid and the sarcophagus itself.

Several theories have been suggested, such as:
- [1933] P. Basiaux: "It's an efficient trap for atmospheric moisture."
- [1934] H. de Varigny: "The opinion of the authorities of the Abbey is that it is due to a natural phenomenon, spontaneous condensation, which is explained by physics at some level"; for that author, "this supposedly miraculous phenomenon is readily explained by physics."

- [1957] Réné Colas: "In that place there are unusual circumstances: exposure to the north in a deep courtyard that gets no sunshine, architectural surroundings of massive construction probably creating thermal currents and in particular good circulation of warm, humid air flowing down from above the southern rock wall, chilling again when it reaches ground level, and depositing its condensed vapor in the receptacle of the sarcophagus."
- [1959] Delaunay-Delapierre and Delapierre-Devinoux [these names have to do with stones and diviners, and they generally indicate that someone was fooling around with pseudonyms]: "The phenomenon is the result of the condensation of vapor from the air."
- [1959] Dupasquier: "The yield on this sarcophagus is very much elevated compared to the setup at Theodosia or Montpellier."
- [1960] Nicolas Chtechapov: "Production of water by condensation is possible here, but it remains doubtful nonetheless."

From Theodosia to Trans-en-Provence

Some people believe that the explanation of the source of the water inside the "holy tomb" is an effect of condensation. This makes it necessary for us to take a short trip, a voyage from a land of olive trees and cicadas all the way to the Crimea.

We start in southern France, stopping at a town called Trans-en-Provence. In this town, the most famous sight is undoubtedly the "air well." This structure reflects an enormous effort to harness atmospheric water vapor to produce water by condensation. It might be relevant to give a bit more information about this.

The air well was invented in 1928, and this prototype was completed in 1931. It was the idea of Achille Knapen, a Belgian engineer. As the next photo shows, the structure is most impressive: it is thirty-six feet around at the base, nearly forty feet high, and it has a stone exterior seven and a half feet thick, punctuated by holes allowing air to circulate. Inside, more than three thousand little slate panels are arranged to maximize the surface area for condensation. Unfortunately, this building never fulfilled the hopes of its creator, who was even granted a U.S. patent in 1931, number 1,816,592, for his efforts. Achille Knapen hoped that the air well could pro-

© H. Broch

duce thirty to forty cubic yards of water per day, but the air well couldn't deliver more than several quarts at best.

This spectacular attempt to recover water from the air is not unique. In certain very humid areas, there are efforts to collect droplets suspended in fog by having them condense on woven networks of fibers, on canvas, in cisterns, or on trees. But we are restricting our attention here to air wells in dry regions, and there is another example. Another water collector was invented in 1929 by Léon Chaptal, director of the bioclimatology agricultural research station in Montpellier. It consisted of a pyramid of calcareous stones, occupying thirteen cubic yards, set up on a concrete platform shaped to allow the recovery of the water. This condenser did actually work, but the average daily water production varied between 0.2 and 0.5 quarts!

These attempts have their origins in the "fact" often cited as *the* proof that such systems could be a practical reality: It is said that the town of Theodosia in the Crimea, four centuries before our era, obtained its water supply from vapor condensers consisting of big heaps of stones. It's true that, at the end of the nineteenth century, the engineer F. Zibold (who was in charge of the Theodosia water supply at the time) made an intriguing discovery: on the surrounding hills were gigantic stone cones, next to the canals that brought water into the city. He was convinced that these pyramidal structures were condensers for water vapor. Each one, according to his calculations, could provide 55,400 quarts of drinking water daily to the ancient city.

Zibold attempted to replicate the ancient accomplishment on a similar scale but was unable to verify his hypothesis. He began the attempt in 1905 and used two thousand tons of pebbles, piled together in a truncated cone 60 feet in diameter at the bottom, 25 feet in diameter at the top, and 18 feet tall. Apparently, it did not yield the hoped-for results. Thus, what had been only a hypothesis quickly became an article of faith.

In short, what happened there? As Pierre Descroix showed decades ago, the figures Zibold published for the amount of water supplied to Theodosia (by the thirteen pyramidal condensers) were unrealistic. Production of the stated amount of water by condensation would involve an increase in temperature of 99°C for all the stone pyramids. This would obviously not be a viable system. As Descroix says, "Taking efficiency considerations into account, we can state that the theoretical maximum yield must not have exceeded 5% of the figure set forth by Zibold." Descroix published his analysis in the journal *L'Eau* in August 1951, in an article titled, "La Récupération de l'humidité atmosphérique" (pp. 127–29).

Two expeditions in 1993 and 1994 were designed to solve the mystery of the air wells of Theodosia and ended by brushing aside the veil of uncertainty. Daniel Beysens was the director of the missions and published his report in *La Recherche* in May 1996 (D. Beysens, A. Gioda, E. Katiouchine, et al., "Les Puits de rosée, un rêve remis à flot"). Beysens's excavations showed that the water mains of the city water supply network (which were, in fact, both medieval and modern) were built near the pyramids "in a completely independent fashion" and that the famous "pyramidal condensers" ringing Theodosia were in fact *kourganes*—Scythian or Greek tombs!

The moisture in the air is certainly recoverable by condensers, but an effective system—let alone an optimal one—is far from what was seen at Theodosia, Trans-en-Provence, or Montpellier. And, although we don't want to upset the parapsychologists enamored of the condensation theory, we must say that the sarcophagus at Arles-sur-Tech can give nothing close to the yield of the ultra-efficient air wells.

The Celestial Origin of the Water, Confirmed

The main result described in the 1961 publication about the sarcophagus at Arles-sur-Tech was confirmed by a recent study that was basically con-

cerned with the amount of water produced by condensation inside the sarcophagus. The results of the study were published in 2001 in the scientific
journal *Atmospheric Research,* in an article entitled, "Water Production in
an Ancient Sarcophagus at Arles-sur-Tech (France)," written by Daniel Beysens and others (vol. 57, pp. 201–12). The result was clear: the water is
certainly of natural origin. The authors of the new study collected data over
nearly three years, in a totally passive manner, meaning that they did not
interfere with the phenomenon. They set up four thermocouples to measure the temperature at four locations: in the outside air, in the water in the
sarcophagus, at the outside stone surface of the sarcophagus, and at the inside stone surface. Measurements were recorded for nearly seven months,
and then measurements of the temperature of the air inside the sarcophagus were added. Additional parameters were measured using two outside
workstations located about 200 and 250 yards from the tomb: amount of
rain, ambient air temperature, relative humidity, and barometric pressure.

The conclusions of the researchers couldn't be clearer, given the type
of investigation being conducted. They clarify—and quantify—the relative
contributions of the two phenomena related to the filling of the holy tomb:
rainwater and the condensation of dew. Evaporation must be considered
as a complementary negative factor, obviously.

To present the exact remarks of the authors, we include excerpts from
Atmospheric Research:

> It has been claimed, since at least the sixteenth century, that a closed
> sarcophagus, located in the courtyard of the abbey of Arles-sur-Tech
> (France), produces hundreds of liters per year. Several hypotheses have
> been advanced to explain this mystery. After about three years of col-
> lecting data, we conclude that the production of water, which amounts
> to about two hundred liters per year, results from a balance among *rain-
> water, condensation* of dew, and *evaporation.* Imperfect closure of the sar-
> cophagus permits exchanges with the atmosphere. *Condensation* is
> about six times more important than is evaporation and accounts for
> about *10% of the overall production of water.* [Italics added.]

Similarly, the article concludes:

The phenomenon of the buildup of water in the sarcophagus of Arles-sur-Tech may be understood as an equilibrium between the rainwater, which slips in between the body of the sarcophagus and the cover, the condensation of dew, and its complementary phenomenon, evaporation. A total annual production of two hundred liters of water has been measured, with dew contributing about 10% of the total production, about twenty liters per year.

As to the analysis of the data, the authors write, "We first tried to correlate rainfall with the rate of production of water in the tomb." From the values obtained, they conclude, "It is clear that a strong correlation exists between these two quantities."

In other words, Beysens and his collaborators established a relationship between the amount of water appearing in the sarcophagus and the amount of rainfall, as did their predecessors in 1961. In addition, they provided evidence on the contribution of the condensation factor.

Before leaving the subject of this new study, we must note that Beysens and his colleagues give a strange assessment of the 1961 work. They write that the conclusion of their predecessors—porous cover, five days' journey to get through it—is without foundation, but they don't support this allegation at all. Indeed, they are wrong on two essential points.

The first erroneous point is that the sarcophagus was "empty" at the start of the measurements. They state that the delay between the rain and the change in the level of water in the sarcophagus observed in the 1961 measurements was caused by a "dead space" due to the tilting of the sarcophagus with respect to the horizontal plane. When the sarcophagus fills, obviously starting with the dead space, you can't get a reading on the water level, since it hasn't reached the apparatus yet, causing a gap. But to use the dead space as an argument to explain the gap in the readings of water level necessarily involves the assumption that the sarcophagus was empty at the start of the measurements. And this assumption is false! The water level was known and perfectly measurable from the very start of the experiments, a fact that can be gleaned directly from reading the 1961 paper. In fact, a simple rule in hydrology even permits estimation of the water level in the sarcophagus at the outset of measurements: it was about 120 quarts! For an "empty" sarcophagus, that's kind of a lot!

The second point concerns porosity and nonporosity. The opinion of Beysens and his colleagues that the marble couldn't have a porous interior is not supported by any argument or by any experiment. It is actually contradictory to the experiments made in 1961 that demonstrated the porosity of the marble cover.

This example shows us that it is not correct to count *facts* and *opinions* as equals, especially when the opinions are not supported by any demonstration or any justification better than "it *seemed natural* to deduce that the sarcophagus was empty" or even "we *thought* that the marble did not have a porous interior."

In Conclusion

With the publication of the 1961 and 2001 papers, information on the famous holy tomb that fills with (now nonmysterious) water has become publicly available. To give a corresponding explanation for the conservation of water in the bottom of the sarcophagus, we would like to call the reader's attention to an essential but often ignored factor: human intervention. For example: "In 1848, as water began to leak out little by little through a small fissure in the bottom part of the tomb, 0.75 meters of water [27 inches] was removed so that the crack could be sealed more easily" (O. Leroy, *Vie intellectuelle* 10, no. 43 [1936]: 194).

We hope that, in the near future, a new plaque will be placed on the fence at the entrance to the courtyard of the holy tomb at Arles-sur-Tech and that the media will not fail to broadcast its inscription: "The holy tomb has given up its secret: rain and condensation."

Radioactivity, or the Devil Gets into the Font of Holy Water

A "natural" phenomenon, like water condensing in the sarcophagus just discussed, that's acceptable. An "artificial" phenomenon, well, that's bad. This type of reaction is pretty close to what we often hear: "that's chemical," it's "artificial," and so on. "Natural" radioactivity, well, okay; "artificial" radioactivity, we're in deep trouble! Radioactivity these days is the subject of an absurd demonization.

Artificially radioactive objects originated in the twentieth century and were quickly incorporated into many human activities. They are used in applications from biological research to the carbon dating of the shroud of Turin (made in 1325 ± 65 years). They have resulted in nuclear medicine and the vaporization of cities, the targeting of prostate cancer cells and a dangerous electricity-generating system, exalted scientific reputations and fierce electoral battles.

The twentieth century saw fearsome cataclysms, in which human fury was unleashed in the most bestial fashion: ferocious wars among the great industrial powers; concentration camps and gulags in which scores of millions of human beings perished on account of their race, ethnicity, religion, or membership in a hated social group. Pitiless colonial wars involved confrontation between those who wanted to preserve an order established by centuries of domination and those who wished to gain independence but professed ideologies often resulting in regimes worse for the people than those they wished to supplant.

According to historians, more than 100 million people were killed by such events in the twentieth century; nuclear arms were the cause of a few thousandths of these deaths. Fear of this weaponry, however, dominated the political thought of many people and influenced the future of nuclear energy, for sometimes legitimate reasons.

Of course, during the cold war involving the confrontation of two gigantic blocs, the fear of a nuclear cataclysm no doubt dissuaded both sides from going over the line. For many in the West, the prospect of a nuclear war wasn't as frightening as the prospect of the military triumph of the Soviet bloc. No doubt the vision of the cold-blooded elimination of certain social classes, such as well-off farmers, the middle class, the intelligentsia, religious groups, and the like, contributed to this preference for death rather than domination by a tyrant such as the one who grabbed the title of hereditary leader of the North Korean people after the Korean War.

On the other side from the Soviets, the precept that "the enemy of my enemy is my friend" led to support by Western democracies of fascist dregs in some countries, like the Greek colonels, Pinochet, and Mobutu. Alas, the list is very long! And the political leaders who were a party to shameful episodes at one point or another during their careers—having to do with colonial wars, Stalinism, Cambodia, whatever—hope that the new century dawning will erase the past, clean their slates, and allow them to start, lily white, on new adventures.

Few of our old political leaders can boast of having had exemplary moral clarity, and this political taint gives a contrasting, innocent quality to politicians now coming of age. Also, the latter were brought up free of the fear of nuclear war.

Remember that, when the Soviet Union collapsed in 1989, the two sides had close to forty thousand nuclear warheads, *each* of which had about seven times the power of the bomb that annihilated Hiroshima. Knowing this, it's not at all surprising that the sigh of relief at the end of the cold war was accompanied by doubts concerning the ability of politicians to manage the enormous power that the development of science puts in their hands.

The nuclear weaponry available is enough, in fact, to melt down almost the entire inhabitable surface of the globe. We know today that among the targets of American nuclear warheads was the country house of the mistress of the second-in-command of an important military organization on the other side. This has got to be the most perfect illustration of the powerlessness of the political classes in the face of the military-industrial complex that was so wisely denounced by Eisenhower.

It is somewhat—but not completely—comforting that friendly conversations between the United States and Russia are aimed at reducing the arsenal to a modest twelve thousand nuclear warheads. This is a focus of concern that ought to mobilize both supporters and adversaries of peaceful or even military nuclear energy. And the solution is infinitely easier to find than the solution to the problem of how to supply energy to the nine billion people of the twenty-first century. They will want to live like us, enjoy good nutrition, have cities lit by electricity, drive nice cars, and go to "rave parties." All of which will lead us to a notable increase in energy consumption and force us to take a dispassionate look at the resources we are leaving to our children. There's not much choice. We must make decisions whose effects will be important in a half-century or six centuries or a hundred centuries. And it would be irresponsible to have crucial decisions depend on short-term electoral issues and the circus of rhetoric and demagoguery that that would imply.

All forms of energy have their own limitations or risks, which need to be assessed. It is essential that we be aware of the natural radiation we receive so that we can judge appropriately the risks of using nuclear power.

Weak Sources of Natural Radiation

The recollection of sudden accidents quickly fades from the memories of those who were not directly affected, no matter the number of victims. Yet a daily threat, no matter how small, is considered intolerable if it concerns an accident whose effects have a continuing effect over many years and involve a huge fraction of the people on Earth. Such threats give rise to generalized emotional reactions.

After the accident at Chernobyl, which spread a massive amount of radioactive material over a large part of the earth's surface, millions of people felt threatened. It is possible that some tens of thousands of them got or will get cancer as a result of this accident. The accident spread the fear of an almost irreversible radioactive contamination of the planet for thousands of years to come, although in this particular situation the long-term global contamination was just 3% of what was released in nuclear arms tests in the atmosphere that basically stopped in 1963. The amount of iodine-131 released by the Chernobyl incident was 0.1% compared to those nuclear tests, although the figure was 8% for cesium-137, which does contribute to persistent contamination of the environment.

The debate on specific pollutants caused by the nuclear industry must include serious consideration of the effects of radioactive substances with long half-lives. After all, these substances must be removed or treated, buried, and prevented from escaping to the earth's surface for thousands or even millions of years. It seems logical to study natural ionizing radiation to see whether it is reasonable to set lower limits for various forms of radioactivity on the planet as a whole. The risks of alternative energy sources must also be taken into account; sometimes these risks can be substantial, although dyed-in-the-wool nuclear-phobes refuse to see them.

Ever since life began on the planet, it has been bathed by a flow of radiation. Such radiation comes from cosmic rays and from radioactive elements in rocks. Added to natural sources are contributions from medical radiology and certain industrial and military activities.

Particles possessing very high energies strike the earth. In the upper levels of the atmosphere, they produce nuclear reactions, which in turn create a huge variety of particles. Except for muons and neutrinos, most of these are absorbed by the air before reaching sea level. Neutrinos are massless and chargeless, and they can go through the earth with less than a chance in a billion of hitting another particle. Physicists have built gigantic detectors in underground caves, and they detected a gust of neutrinos in 1987 because a supernova blew itself to bits 150,000 light-years from Earth and released a tremendous number of neutrinos during the explosion.

On the other hand, muons are charged and interact with matter, notably with the human body: on average, five muons go through you each second. And a muon releases into the body about five times as much energy as an electron emitted by an ingested source of radioactivity.

At high altitudes, radiation is much more intense than at sea level. On mountaintops, there are electrons, gamma rays (that is, high-energy X-rays), and muons; when you are at the altitude where the *Concorde* flew, you can add in protons, neutrons, and pions.

The diagram shows that, when a very-high-energy proton enters the atmosphere, it produces a cascade of exotically named particles, indicated by Greek letters like μ, ν, and γ, in specific families like electrons and neutrons. Because of the huge diversity of particles, the Greek letters have gotten used up, and people have started to draw upon the Hebrew alphabet for names.

Neutrons released into the atmosphere by nuclear reactions interact with nitrogen in the air and produce carbon-14. This is a radioactive form; carbon-12 is the common, stable form. The half-life of carbon-14 is thousands of years, and we inhale it, just as every living organism does, plants and animals alike. Ten percent of the interactions between released neutrons and the atmosphere also produce tritium, equally radioactive. Carbon-14 is commonly used to date any objects containing carbon, such as matter derived from living things, like bones or charcoal. This is done by measuring the ratio of carbon-14 to carbon-12. The ratio permits estimation of the year when the carbon-14 was inhaled as carbon dioxide by any plant or animal. Its concentration from that point decreases because of radioactive decay once intake from the atmosphere has ceased.

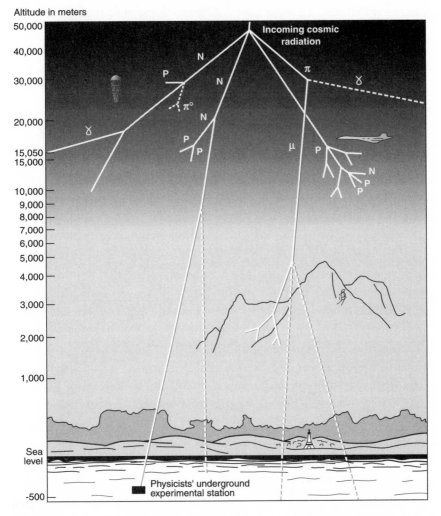

Altitude in meters

AGRAPH

Natural radiation is also emitted by radioactive elements in rocks and living tissues. Potassium-40, the radioactive isotope of potassium, makes up 1/10,000th of naturally occurring potassium atoms. Its half-life is 1.3 billion years. Like radioactive uranium-235, it was originally produced when the earth was formed, and it is still around today. It was in the dust of dead stars whose agglomeration gave birth to our solar system. Potassium has an affinity for most types of living tissue, and thus the radioactive form is always present in living things. Other elements also have long-lived radioactive isotopes. Thus, a 150-pound human being has radioactive material giving rise to 10,000 nuclear particle emissions per second, a small fraction of which are detectable, while the majority end up being absorbed within the tissues.

Natural radiation, then, is made up of muons from cosmic rays, as well as beta and gamma rays given off by rocks in varying amounts; these constitute the background radiation in the midst of which life has developed. It is true that background radiation may have caused genetic damage, but in the course of evolution, living tissue has developed mechanisms for gene repair.

It is simply ridiculous to demonize radioactivity at the level of natural radiation, if people aren't taking protective measures against slightly radioactive construction materials or cosmic rays. And that would be really paranoid! From the point of view of society, it is always advantageous to reduce radiation levels if the cost is less than the expected benefit. But we don't know whether this is true today, one way or the other.

The authors have met a Swiss professor of physics who was motivated by an antinuclear fervor disguised as ecological concern. He criticized one of the world's leading physics research institutions, the Centre Européen de Recherche Nucléaire, for saying that their research activities gave off extremely low levels of radiation, much less than natural background radiation, and for claiming that their activities therefore posed no risk. He held the view that, at high doses, organisms mobilize defense mechanisms against radiation, while at very small doses, the attack is insidious, undetected, and thus more harmful. If we agreed with him, which we certainly do not, we would be able to say that, due to omnipresent natural radiation, we are always above the danger threshold and in the zone of innocuous radiation. Receiving a tiny fraction of natural radiation has no importance, except when the antinuclear ideology of the irradiated person causes an enormous amplification of the effect, as was the case for that professor.

Variability in natural radiation levels due to geography and altitude is enough to rule out taking such professors seriously. Yet estimation of the effects of small doses of radiation on the human organism is important to society, and controversial statements about such effects give rise to polemical debates. Thus, it is important to understand the relative importance of all sources of radiation and their effects on health.

The field of physics offers instrumentation that is very sensitive to radiation. Physicists can easily detect the passage of a single rapid electron emitted by a radioactive atom. What we're interested in here is the sensitivity of the human body. The body is made up of nearly 10^{28} atoms (that is, the enormous number denoted by a 1 followed by 28 zeros), and they are divided up among 10^{14} cells. Thus, each cell has 10^{14} atoms in it, of which 10^5 are perturbed by a fast-moving electron making its way through. We've got to realize that, during a lifetime, each cell accumulates millions of spontaneous lesions, which are constantly being repaired, more or less effectively, by marvelously complex mechanisms that living tissues all possess. The effects of these microlesions caused by various types of rays are hard to understand.

Most particles, when they react with the atoms of the human body, snatch out electrons from those atoms, whence their name *ionizing radiation*. They have roughly the same effect on living molecules. Differences exist among particles with respect to their capacity to destroy living cells. These differences occur because radiation may be made up of heavy, slow-moving particles such as alpha particles (which are helium nuclei) or light and fast-moving particles like electrons and gamma rays. Biologists assign a coefficient Q to represent the biological effect of radiation used in diagnostic and therapeutic radiology. For example, if the value 1 corresponds to the power of X-rays, gamma rays, and rapid electrons, it would be 10 for protons or rapid neutrons, and 20 for alpha particles.

Some Units of Measure for Radiation

1 Bq: The becquerel is the amount of radioactivity in a source undergoing one atomic disintegration per second.

1 Ci: The curie is the amount of radioactivity in a source that undergoes disintegrations 3.72×10^{10} times per second. This is the radiation given off by one gram of radium.

The physical effect of ionizing radiation is measured by the amount of energy tranferred to a kilogram of living tissue.

1 Sv: The sievert is equivalent to one joule per kilogram. One joule raises the temperature of a gram of water by 0.24°C; 1 mSv = a thousand sieverts.

The biological effect of radiation is related to the physical effect by the coefficient of biological effect, Q. The formerly used unit of irradiation was the roentgen, the physical dose received at 1 meter's distance from a 1 Ci source of radiation in an hour.

The biological dose (rem, for roentgen equivalent in man) was related to the rad (radiation absorbed dose) as follows:

1 rem = 1 rad × Q
1 Gy = 1 gray = 100 rad
1 Sv = 1 Gy × Q
1 rad = 10 mSv

The "dose rate" is the dose per unit time.

Source: Georges Charpak and Richard L. Garwin, *Bulletin de l'Académie nationale de médecine* 185, no. 6 (2001): 1087–96.

Dari: A Measure of Annual Dose Due to Internal Radiation

Two physicists proposed a new measure of the effects of radiation in the wake of discussions that followed the publication of their book, *Feux follets et champignons nucléaires* (by Georges Charpak and Richard L. Garwin, published in Paris by Odile Jacob, 1997). They suggested that specialized nuclear agencies should adopt the dari, from a French acronym meaning annual dose due to internal radiation. The measure would be meaningful to anybody willing to think a bit about the subject and would protect thinking people from any demagoguery about the topic of radiation.

The very thing that makes radioactive materials powerful research tools—the exquisite sensitivity with which they can be detected—is turned against them as an indication of their danger. One can measure the disintegration of a single atom (1 becquerel) of radioactivity with a detector, but usually billions of atoms are needed to measure the dose received by nonradioactive objects. So you get people condemning radioactive contamination measured in becquerels, while seemingly ignoring the 10,000 becquerels permanently lodged in our bodies and obscuring the relative effect of such contamination compared with the infinite number of assaults on the body from lifelong internal sources or due to human activity.

Since the human body ordinarily sustains the effects of radiation given off by potassium-40 and carbon-14, which are always there, it has been suggested that people use, as a reference unit of measure, the effect of natural radiation. The intensity of this background radiation is about 10,000 becquerels.

A 500-dari limit is imposed on workers in the nuclear industry in some countries; such an exposure corresponds to an estimated reduction in life expectancy equal to that produced by smoking ten cigarettes per month. This should be compared with specific risks associated with various professional occupations rather than being considered in isolation. After all, even driving a car produces a greater risk, as a result of the carcinogenic nature of the exhaust.

Recent accidents involving release of radioactivity at nuclear power plants have engendered controversy, but the effect of such releases is generally less than a tenth of a dari. It is important to keep matters in perspective when sounding the alarm about contamination incidents whose effects may be considered nonexistent on the scale of human life.

An important and real problem for nuclear energy is the management of nuclear wastes from spent fuel. Certain radioactive elements have long half-lives, on the order of hundreds of thousands of years. People have come up with a variety of ways to render these elements harmless, for example, burying them deep in underground caverns in desert regions. In China and Mexico there are deserts where it hasn't rained for two million years. The burial of the world's nuclear wastes in deep pits in such areas would constitute a considerable source of revenue for those countries. The possibility of conditioning these wastes by fusing them in glass and then sealing them in a thick, metal container is under study. Some people think that burial should be close to the surface so that the containers can be retrieved, permitting the eventual destruction of the wastes using nuclear transmutation methods. Others want to have the wastes buried in very deep caverns in containers designed to withstand thousands of years of irradiation. The governments of several nations are studying what will work best; the French government, in particular, is committed to making a decision in 2006 on how it will proceed in the matter, based upon data that emerge from relevant experiments.

The engineers must meet one criterion: the dose of radiation that escapes cannot exceed 2% of natural background radiation, keeping in mind that natural variation may be 250%. Radiation emitted by these wastes in keeping with such limits is acceptable.

Some people start off with the a priori requirement that no radiation escape at all, recommending a complete halt to the exploitation of nuclear energy. But natural radioactivity alone is of an order of magnitude that would be produced by a colossal nuclear industry, so absolutist objections of that nature might be characterized as a sign of mental rigidity. In the oceans alone, there is enough naturally occurring uranium to inspire some to consider its extraction, which would take care of fueling nuclear power plants for millions of years. And this uranium poses no danger. You may fearlessly bathe in the ocean, as long as you are careful to avoid the skin cancer caused by exposure to the sun, a risk factor that you don't hear activist groups denouncing.

We really should concern ourselves with actual dangers—they are bad enough. Some of them take us by surprise. For example, the events of September 11, 2001, in New York and Washington revealed the danger of certain forms of terrorism. We have to face up to them and protect all dangerous sites, whether nuclear or not. A careful assessment of the vulnerability of many industrial locations, like chemical factories, is also of the first importance. Remember, the Union Carbide disaster in Bhopal, India, snuffed out the lives of six thousand people in an instant. Those responsible for deciding what sources of energy will be used in our society must consider and compare all available potential resources, without yielding to the demagoguery of the noisiest groups or those most richly endowed.

The misery caused by overpopulation wreaks a lot more havoc than accidents can, whether nuclear or not. Overpopulation creates desperate masses with nothing to lose and no future in sight, an ideal breeding ground for terrorism, which will also gain from the inevitable progress of science and technology. Such progress is especially conducive to the production of low-cost weapons of mass destruction, like bacteriological weapons. There is only one antidote to these problems: solidarity with those who didn't have the privilege of being born on the right side of the border, inside a prosperous country with stable institutions, where all the people have the illusion that their future is assured if they conduct themselves like good little consumers. After all, it's by sheer, dumb luck that Swiss postal workers enjoy a standard of living vastly better than that of the Italian and even better than that of the Indian workers, since none of them contribute in any significant way to the underlying differences between their countries.

This solidarity goes hand in hand with provision of aid for development. The laws of the marketplace alone will not work miracles. Solidarity and aid may have an effect, given an enormous investment in education, the sole antidote to the influence of religious or political fundamentalists, who foist their fears on the directionless, uninformed, miserable masses.

A Funny Thesis

Elizabeth Teissier earned her doctorate in sociology from the Sorbonne with a thesis extolling the virtues of astrology. A detailed analysis of her thesis may be found on the website of the University of Nice–Sophia Antipolis, at www.unice.fr/zetetique/articles/articles.html. The awarding of this doctorate created quite a stir because it demonstrated how professors could allow waves of muddy thinking to lap at a prestigious Parisian university under the pretext of a laudable openness of mind. In a democratic lay republic like France, it is certainly legitimate to let all the flowers of human creativity blossom. No one is offended by the teachings and musings of theology faculty. Perhaps we do lack university departments of witchcraft, occult sciences, astrology, and other activities in which the talents of tens of thousands of professionals flourish—professionals who offer advice, care, and consolation to millions. With university departments to deal with such subjects, we could honor these people with the title of doctor, and the most talented among them could run the place as rectors and provosts! It really is essential to avoid having these subjects enter the university setting other than as subjects for study in sociology, psychology, or psychiatry.

We have a recommendation for students wishing to undertake a doctoral thesis in sociology: avoid the advisers and examiners who made up the doctoral committee for Elizabeth Teissier, Ph.D. People will not take seriously the contents of any future thesis with their names on it. And with their reputation goes that of the university. Can a thesis be declared invalid? Maybe not, but we would seriously like to see a thesis entitled, "An Analysis of the Psychosocial Factors Leading Eminent University Professors to Grant the Status of Science to Primitive Superstitions."

Chapter 4

The Right to Dreams and Clarity

A Knee-jerk Reaction

It makes you wonder: before World War II, a French high school teacher in Casablanca, obsessed by dowsing, claimed to mark his test papers by following the instructions of his pendulum. Then you breathe a sigh of relief when you learn that the administration took steps to prevent him from using his special powers in the educational system. And you tell yourself that was a long time ago.

Then you are stunned to learn that a newspaper article in late 1991 reported that a big-city high school graduating class demonstrated its initiative, as well as the growing role of alternative sciences in business, by organizing a meeting to display the merits of numerologists, astrologers, and their ilk. And you just can't believe your eyes when you read that the meeting's closing reception, presided over by the principal of the local high school, drew a large crowd.

Then, when more than a dozen of the faculty sent this article to one of us to share their amazement, we thought it our duty to transmit the resultant feelings of incredulity. Our communication took the form of a letter to the principal expressing the inevitable disappointment felt when education officials lend their support to the "dumbing-down" of our culture, at a time when there is so much talk about the need to improve science education. When reading such articles, we "feel ashamed of the behavior of these people with whom we are supposed to share common pedagogical objectives."

This knee-jerk reaction on our part drew an unintentionally laughable reply from the principal's office. The reply stated that the newspaper was erroneous in reporting that the closing reception took place under the aegis of the principal. Regrettably, the hundreds of thousands of readers of the newspaper weren't among those informed of this. But the best part of the reply was the following passage: "I point out that the heads of institutions, being autonomous, are free to organize any demonstration, meeting, or colloquium whatsoever that may seem suitable to the conduct of their establishment." We informed this principal that the pedagogical objectives that we have the honor to share with many (but obviously not all) colleagues would preclude an "astrology day" or "numerology day" or any other pseudoscience being taught in our colleges, high schools, or universities. The incident illustrates the paucity of knowledge and information possessed by some (fortunately rare) educators in positions of authority in the educational system.

From Artisans to Multinationals

There really is a problem. At every level, society is infected with unscientific thinking, with potentially disastrous consequences. In the space of a few years, the occult has gone from an individual, local craft to an international big business. This growth gives rise to the various excesses and trends we've been discussing and must be stopped because, once belief in the supernatural reaches a certain point, the seriousness of the damage may grow at an accelerating rate.

As we've already emphasized, you can't lead people to accept the grossest errors, the biggest falsehoods, and the least justifiable reasoning without consequences. If you lead people to accept nonsense, you cannot do so without seriously inhibiting their intellectual development, without making them doubt the validity of science and all the values that come from reason. Our era is unfortunately a great example of this because of the magnifying power that the media provide the pseudosciences.

The reemergence of occult, paranormal, or magical practices has been oddly swift—so rapid, in fact, that one must ask oneself this question: What are the favorable circumstances that have created such a need and have favored, perhaps unwittingly, its growth? For one thing, financial stakes have a great effect on this trend. But maybe the problem is more serious than that. The geneticist Albert Jacquard said it very clearly in his foreword to *Incroyable . . . mais faux!* (by Alain Cuniot, www.book-e-book .com, 1989): "To transform citizens into passive sheep is the great dream of the powerful. There are many means to this end; poisoning their minds with pseudoscience can be very effective."

How else can we understand it when a public television network devoted to culture, which is otherwise distinguished by some excellent programs, set the tone for its series, "Science and Technology," by offering a study of the pyramid of Kheops as its premiere broadcast in 1992? Not the marvelous pyramid itself, but reduced-scale models of the pyramid, presented as wave concentrators with which one can mummify meat. They can also be used to resharpen used razor blades because (quoting verbatim here), "metals undergo changes inside the pyramid." All this from pseudoscientists in white lab coats.

One would have expected that the storm of reactions to this odd initial broadcast would have made the network management more careful, but nothing could be further from the truth, in light of more recent examples. On June 8, 2001, we saw a lengthy show on the Arts channel in France, concerned with paranormal phenomena, presented with almost no critical or skeptical point of view but a profusion of positive assertions concerning the quest for ghosts and poltergeists.

This sort of thing shows us how media that are excellent in other ways, in cultural domains, can be rotted away by the infection of unclear thinking.

Rationality and the Construction of Beliefs

Given that rationality has the first and foremost place in the evaluation of our beliefs and in their analysis, does it enjoy the same preeminence in the *construction* of our beliefs? As surprising as it may seem in our scientifically and technologically oriented society, the level of belief in superstitions is directly, not inversely, associated with educational level. But it is not educational level that determines the presence or absence of particular beliefs: rather, given belief in a certain "paranormal" phenomenon, the educational level drives the choice of the *form* of belief most compatible with that level.

For example, take the case of dowsing: The pendulum is the symbol of that pseudoscience. It can be used to perform divination, to identify a playing card that's been selected, and for lots of other applications. For a person with a higher educational level, the same pendulum might be used instead to study the effects of the earth's energy waves on living things or magnetic variations indicating underground water, studies that—strangely enough—have been defended by a capable physicist.

The Rise of the Occult

In France, the country of Voltaire and Condorcet, a nation of skepticism and the Enlightenment, occult beliefs flourish! We can demonstrate this with some data (see *Skeptical Inquirer,* May/June 2000, 34–39).

Survey of Undergraduate Science Students, University of Nice, 1982–1983

Percentage who think it's	Using the mind to bend spoons	Relativistic time dilation
scientifically proven	68%	18%
generally recognized as plausible	14	18
unlikely	15	7
pure theoretical speculation	0	52
completely disproven	3	5

The table summarizes the results of a survey carried out twenty years ago by one of the authors. The survey concerns the beliefs of beginning undergraduate science students with respect to psychokinesis and relativity.

Psychokinesis or telekinesis is the movement of objects at a distance without any physical contact, using only the power of mind over matter. This principle is exemplified by the bending of spoons using only the mind; in the 1980s, Uri Geller was a fashionable practitioner with many articles and television programs devoted to his spoon-bending. Nearly seven in ten students considered psychokinesis scientifically proven, while half thought that relativistic dilation of time, the physical manifestation of relativity, was pure theoretical speculation!

A recent example of the deterioration in people's ability to distinguish between science and pseudoscience came to us in the form of an e-mail received from Nicolas Hergott, a student of physics at the University of Toulouse, in February 2001. He had discovered a 1999 article in the *Bulletin of the Society of Physicists* on occult beliefs in the teaching profession, and he wanted to add his own "modest (but scandalous) contribution to statistics."

During a meal, he asked eight of his fellow students if they believed in certain paranormal phenomena. The results were so shocking that he declared himself "ready to dedicate his entire career to combating this mental weakness." Among the eight, one regularly went to see a "specialist in magnetic fields," and this surprised only one of the others. Three believed in psychokinesis. Four believed that certain observable phenomena would remain unexplained forever and couldn't imagine how to dissociate some of these phenomena from concepts like the soul or God. Every one of them, without exception, believed in telepathy. He added that it is "impossible to discuss the believability of these phenomena without being extremely annoying to these believers."

This particular, limited example is significant, and we don't consider it anecdotal (since the problem is widespread). It concerns future physics teachers and shows that the level of gullibility at the start of the new millennium is higher than you might expect.

Perhaps beliefs in paranormal phenomena, while widespread, are not deeply rooted. If the public sees perfectly regular people on television declaring that they have seen some supernatural phenomenon, sometimes authenticated by others whose phony authority is backed up by prestigious diplomas or a high social position, it can be hard to keep enough distance to say, "They are trying to fool me, they are naive people who are lying, or lying to themselves, because they lack a critical mind." Only those who have staged something themselves can really know how far human credulity can extend.

When it comes to phenomena governed by chance, how many people are capable of analyzing a simple fact dispassionately? For example, if ten thousand people watch for a phenomenon with only a one in a thousand chance of happening to a given person, there'd still be nearly a dozen with the rare privilege.

False beliefs based upon unlikely phenomena are so commonplace that special precautions must be taken, for example, by those judging the effects of medications on patients, in order to be sure that accidental findings, amplified by the imagination, do not taint objective interpretation. And lots of distinguished physicists have published erroneous "discoveries" because of accidental fluctuations supporting unlikely theories or confirming ill-founded ideas.

A Paradoxical Situation

Even if false beliefs are undergoing a huge expansion, we should still bear in mind that the phenomena on which the beliefs are founded are not becoming more numerous. On the contrary, the corpus of observed supernatural phenomena is shrinking like a wool sweater in a clothes dryer. As everyone knows, fairies appearing in gardens, Tibetan lamas levitating, ghosts, and witches on brooms are all making themselves scarcer and scarcer.

The magnitude of the phenomena is also decreasing, and this is true for all paranormal phenomena. For simplicity, we'll just consider the example of telekinesis that we discussed before, in connection with the survey. Let's examine the change in the renowned power of psychokinesis—the ability to move objects at a distance using only mental concentration—over the course of time.

This power, sometimes known as "mana," is supposed to have moved the giant Easter Island statues, weighing several tons, several centuries ago. In the 1850s, the same power supposedly moved heavy tables, weighing hundreds of pounds. Several decades later, it was time for poltergeists—the "knocking spirits"—and the movement of casserole dishes and cooking utensils, items weighing a pound or two. In the 1970s psychokinesis could move little objects like chess pieces. These days, the same power allows a medium exerting enormous concentration to move a tiny piece of

Curve showing the mass of objects supposedly movable by psychokinesis, as a function of time. Claims for the same phenomenon become less and less dramatic as technology advances.

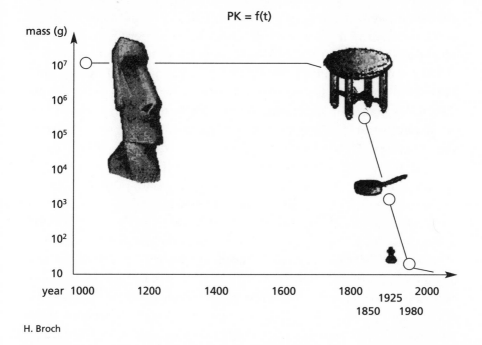

H. Broch

paper, maybe a gram! So, psychokinetic phenomena have declined: as the ability to validate them scientifically has improved over time, they have decreased by a factor of more than a million.

Moreover, the means of verification don't have to be very elaborate to work. For example, the advent of infrared photography (involving the ability to detect objects in the dark) effected a change. Powerful spirits used to manifest themselves around tables in poorly lit circumstances, no doubt because their eyes are very sensitive and the spirits couldn't work in bright light. Since the advent of infrared photography, the spirits have lost their powers.

Thus, the current situation is a bit of a paradox: beliefs in the paranormal are more widespread than ever and are infiltrating the ranks of the educated, while the number and strength of paranormal phenomena are decreasing drastically. The main reasons for this apparent paradox are as follows. First, information replication by electronic media, such as radio, television, and the Internet, was unparalleled in previous generations. At the start of the twentieth century, a village's evil spirit had only a local reputation; nowadays all it takes is an ordinary little poltergeist pestering a Belgian village in order for CNN to present a story that circles the globe.

The second factor in the paradox is lying in the media and "moral drift." The media—okay, the generalization is obviously unfair—have a concept that allows them to disclaim responsibility. They are meeting consumer demand. But they aren't giving their readers, listeners, and viewers what these audiences seek. They are not conductors, intermediaries, or "mediums" of demand: they create a demand and then act as if they are merely responding to it. They are not neutral instruments; on the contrary, they selectively amplify phenomena involving a return to a tawdry, sham religiosity.

Finally, our educational system is hooked up to all this as if to a car's transmission belt, and this, too, contributes to the paradoxical situation. Contrary to what one might have imagined a priori, the educational system does not confer any immunity against superstitions, and sometimes it even becomes a means of transmission for pseudoscience and other nonsense. There are plenty of alarming examples for us to present—from miniature Kheops pyramids accelerating the aging of wine sold by a teachers' cooperative to lists of books, directed at teachers, that praise the merits of magnetic fields or astrology.

Reason and Sensation

Here's a complementary explanation of the apparent paradox concerning the persistence and growth of "paranormal" beliefs and the decline in number and power of paranormal phenomena: it's that we are living right now through an unusual phase of modification of the processes by which knowledge is gained. The current growth of information is in fact essentially, if not solely, characterized by an explosion of visual images with sensory impact and the decline of symbolic writing and reasoned analysis.

As a means of communication, writing permits an analysis that is detailed, organized, critical, and available over a period of time, whereas the new media give prominence to instant images and stimuli. This substitution of "visual image + sensation" for "written symbol + reasoned analysis" results in a progressive and stealthy replacement of reason by sensation. But just because you act on "gut feelings" is no reason to keep you from also working with your brain.

Democracy Imperiled

Science is central to modern culture. Every scientist, being a citizen living in society, can and *must* raise the issues and problems posed by the growth of the pseudosciences and unfounded beliefs. Indeed, the scientist has the obligation to demonstrate that belief in paranormal phenomena is an obstacle to the future of human freedom and human potential.

Lacking free will, human beings would just be objects; is the destiny of human beings to be found in the intricate paths of the planets and stars in the heavens?

If people are mere primates, should we rely on aliens to come to Earth to educate our species, incapable of developing by itself?

If people are but vassals, obedient subjects, are they then condemned to have as masters those mediums with supernatural powers?

Despite what they admit, most astrologers and parapsychologists ultimately do answer the above questions in the affirmative. Everything they suggest is really nothing more than complaisance and a cowardly rejection of the attempt to understand the universe around us.

This reduction of humanity is much in evidence in the techniques that sects use to depersonalize the individual. The method always involves declaring that "forces" can be channeled by certain people (the "elect," the "anointed," the "specially gifted"), while others are worthy only to marvel at them. However, the special people are not themselves the generators of the forces—they only "focus" them; they are only "mediums."

Thus, we are witnessing a mystification of knowledge, which results in a concept of the world in which many things are forever outside the understanding—and the control—of most people. This odd idea implies a stratification of society into two groups: an upper level consisting of the powerful, who know and do everything, and, far beneath them, those who wonder, admire, and follow without understanding. Ultimately, this leads to a complacent fatalism and the loss of individual responsibility.

The scientific outlook and good citizenship both require similar mental and moral soil for their development. A truly democratic society presupposes citizens capable of reflection. This is why the results would be even worse than is generally imagined if the scientific spirit—that is, the critical spirit—were to be supplanted by credulity. Let's never forget that the right to dream cannot attain its full value without the right to think clearly.

Chapter 5

A New Millennium Dawns

To some, science has enabled the human race to flourish by adding the treasure of knowledge to the various attributes that distinguish people from animals. To others, science has been abusive, giving rise to offspring that threaten to become monstrous.

Who wouldn't be sickened by the idea of a world invaded by human clones, mass produced by the same techniques used in chicken farming, indoctrinated from infancy by crazy leaders who can mold their thinking so that they believe that it is lovely to die suicidally for a "good cause"?

Human societies must confront a dizzying acceleration of change, caused by the repercussions of scientific discoveries. Yet they seem powerless to face up to changes threatening their economic, political, cultural, and even physical environments. Rational use of scientific thought to assess and cope with real or imaginary dangers is sometimes made impossible by habits, beliefs, and superstitions in which people have been immersed since time immemorial.

We have been driven to touch on a wide variety of subjects in this book. These include threats hanging over the human species because of the vertiginous acceleration of change due to scientific advances and the gravity of disruptions to the physical continuity of life on the planet. They also include the effects of the most widespread superstitions seen in ages and their political exploitation for very short-term objectives, preventing the use of science itself to evaluate and escape dangers.

It has become evident to us that the principal antidote to scientific illiteracy is education. We wanted to make a contribution in this regard by providing an overview of the most widespread superstitions: astrology, "paranormal" beliefs, and other crazy views. We hoped to sharpen the critical faculties of our readers by introducing them to tricks and games that would point out their own credulity.

We are alarmed by the gravity of the problems that are going to beset our children, which they will inherit because of our actions—actions often distinguished by a lack of compassion in relations between societies and a longstanding indifference to environmental concerns.

Human activity, whose effects have been greatly magnified by science, can change the physical characteristics of the earth itself, disturbing the miraculous but delicate equilibrium among the factors that made possible the origin of life and the richness of its evolution. Human activity has led to population growth and the frenetic exploitation of natural resources. The time has inevitably come when we cannot be content to live in our own, more or less abundant niche, consuming our fill of the products of high technology and struggling mightily against obesity, the greatest threat to health for many of us. If we do not learn to live in our niche, however, gigantic explosions and unspeakable suffering await us.

The question is inescapable: Isn't scientific thought the *indispensible* companion to wisdom, to clear thinking, and to the love of those virtues, which is expressed not only in vain incantations to the sky but also in logical actions. Scientific thought will be challenged by a curse, the rising power of obscurantism, which ultimately springs from ignorance, fear, and the superstitions carried along as a relic of the past.

All Ideas Are *Not* Equally Valid

In the year 2000, the French post office created a calendar featuring photographs by Cartier-Bresson, which also presented the astrological predictions of Elizabeth Teissier. Spread over two full pages, with colorful drawings, the predictions of our seer appeared under the title, "What does the year 2000 hold in store for you?" Telephone numbers and an Internet address were provided for readers wishing to know more. The post office delivered these calendars by the millions.

The same calendar had a feature, "Tomorrow: The Twenty-first Century," which also occupied two pages, and not a bit more, even though a vision of a century, not solely one year, was being presented. On those pages we learned about many kinds of robots and automata, a library in a single book, computers and the climatology of the future, the car of the future, mass transport of the future, energy in the twenty-first century, video cameras, aircraft, intelligent highways, and a post office rather less intelligent than the highways, since it is contributing to credulity throughout society.

The "postmodern" movement posits that anything is morally acceptable, that anything is valid, and that any opinion is as good as any other. These ideas, especially the last, permit the development and spread of irrationality and contaminate even science. No, all views are *not* equally valid. For example, if we tell you that, when we let go of a book this morning, the book fell on the ground, you wouldn't need much in the way of proof. If we tell you, however, that when we let go of a book, it went up in the air, contradicting the law of gravity, you must—not "may"—you *must* demand much more compelling evidence than our assertion alone. The burden of proof is always on those who assert something new. The more the new claim lies outside previously established natural laws, the stronger the evidence to support it must be. This is especially true if the claim *contradicts* established laws.

This is why all ideas aren't equally valid. This is why fashionable speculation does not at all merit the scientific label "a working hypothesis," not even when the speculation comes from a person who claims to be a scientist and has diplomas to show for it.

Scientists and Journalists: On the Same Side of the Battle

We aren't fighting against the irrational; rather, we are fighting for something. We are fighting for people to get at least two sides of a story. After all, superstitions don't bother anyone unless they are presented as scientifically proven facts. We start off the new millennium at a new stage in the struggle for reason. And the role of the media will be the key to determining the outcome.

Fortunately, there are journalists worthy of respect and esteem. What shows they make, delighting us, leaving us touched by the tube's revelation of aspects of the physical or artistic or philosophical world that we had not known! What pleasure there is in certain discussions among people on TV, perhaps with opposing views, who don't give the impression that viewers must be presented with predigested content.

You wouldn't ask someone who knows nothing about cars to offer an opinion on a new car's motor, and you wouldn't ask a barber about rare genetic disorders, but when it comes to the sciences, anything goes. A person who couldn't explain to his children why water boils can get on TV and trumpet his expertise on the effects of radioactive krypton coming out of people's chimneys in The Hague. Sure, it is vital to avoid an intellectual dictatorship by a few lobbies, and fierce debate is sometimes desirable, but not when the debate is dominated by ignorant people who act like they believe that coastal currents depositing a little naturally radioactive sand on a beach merit media attention. It seems legitimate to us that certain organizations, like the National Academy of Sciences, should be able to demand the right to reply when gross errors are presented and spread.

In this arena, media people ought to play their role, play it fully, and consider seriously what is meant by the concepts of neutrality and responsibility. Many media figures have, instead, an annoying tendency to hide behind their "necessary" neutrality. Thus, they give us reporting without serious inquiry, "raw information," under the pretext of letting the listener judge for himself or herself. They just forget—or pretend to forget—that the audience's critical faculty is applied to a vacuum if presentations lack information or lack information that is sufficiently objective. These high priests of a new religion are afraid, above all, of attracting the wrath of the great god Ratings. This is the source of the fallacious recourse to "neutrality." And the difference between neutrality and cowardice becomes very small.

That really must change, although another change is also necessary: In our world, the line between reality and virtual reality must become sharper. When scenes on news programs are staged and then exposed as mere propaganda, it shocks the mind, but the same minds that are shocked remain completely indifferent to the same shameful manipulation, the same staging of events presented as reality in certain television broadcasts about paranormal phenomena. Where is journalistic responsibility?

The media are the carriers and transmitters of the virulent germs of the pseudosciences; at the same time, they are the best means of fighting off such infections. The media, thus, must deal with their ambivalence and their responsibilities. All hope is not lost. The media sometimes actually do present an interesting item that makes you think, like the newspaper story we saw in *Nice-Matin* on March 24, 2001. It was about an experiment conducted during Science Week in Great Britain. It was so good! They tested people's ability to predict the stock market, and a four-year-old girl did better than a financial specialist and an astrologer.

It's Never Too Late

It's happened many times that people who acted as mediums admit to fakery late in life. The two girls—Frances Griffiths, age ten, and Elsie Wright, age sixteen—who saw winged fairies fluttering in the air and photographed them admitted their deception when they were seventy-six and eighty-two!

Margaret Fox Kane founded a spiritualist movement around 1847 in New York State. She made a long public confession forty years later, explaining that her movement was nothing but "fraud, hypocrisy, and illusion." The complete text of this confession can be found in H. Broch's work, *Au Coeur de l'extraordinaire,* published by book-e-book.com in 2002.

Even more often, suspicions of fraud arise or fraud becomes plainly evident. The example of the well-known Zener cards is a case in point. The cards come in five types: square, circle, cross, star, or blank; they were developed and sold by the American Rhine Institute and were so poorly made that their backs were not always symmetric despite more than sixty years of fiddling with them by the parapsychologists who developed and distributed them. You can very easily recognize the cards from the back and thus improve the statistics on successful guesses without any clairvoyant powers or any other supposed gift of extrasensory perception.

Although this statement may shock the ears of parapsychologists, fraud is perhaps *the* principal source of successful results in parapsychology. Even Walter J. Levy, the director of the most famous research institute in parapsychology and Joseph Banks Rhine's own successor in Durham, North Carolina, has been caught in the outright commission of fraud. But that doesn't stop his works from being cited to this day as scientific studies that have proved beyond the shadow of a doubt the existence of parapsychological powers!

Powerful Armies Can Let Themselves Be Fooled

During the cold war, the American army took an interest in magic wand experts who claimed to be able to detect Soviet submarines by running their divining rods over maps. Alerted by their own unbeatable intelligence services, the Soviet military asked Ukrainian scientists like Leonid Pliouchtch to meet the challenge. Consider the following lines, written by Pliouchtch about his experience; they illustrate the level of mediocrity to which the Soviet leadership had fallen (which explains to a great extent the collapse of that country). The text is from his book, *The Carnival of History,* published in Paris by Seuil in 1977.

The more I immersed myself in the flood of books on telepathy, the more my interest in paranormal phenomena grew. Together with a group of my colleagues, I went to the chairman of psychology, and we proposed to establish a discussion group concerning telepathy. . . .

I went to read papers on telepathy at various institutes in hopes of attracting various specialists to our group. At the same time, the first articles devoted to the subject appeared in the Soviet press. I learned that a collaborator of the academician Bekterev, B. B. Kajinski, was in Moscow. In the 1920s and 1930s, he had organized experiments on telepathy with Dourov and Bekterev. I wrote to Kajinski and went up to see him in Moscow. He received me very cordially, seeing in me a young man ready to take up the torch and continue the work that he had undertaken in this area before the war. There were four of us at the table: Kajinski, his wife, a young doctor named Naoumov, and me. During the gathering, Naoumov proposed to carry out a pseudotelepathic experiment, asking me to press his foot with mine at certain times. I agreed. During the fake demonstration, Kajinski did his best to discover how it was done, but we succeeded in fooling him and convincing him that he had really been present at an authentic demonstration of telepathy. I felt very ashamed in front of him but couldn't make an issue of the phony situation we had set up.

My interest in Kajinski soon evaporated, but that incident allowed me to develop a principle that has guided me ever since. In every experiment, the scientist involved with parapsychology must assume deception or self-deception and set up the experiment in such a way as to make such falsehood impossible. The parapsychologist cannot take another at his word.

This type of fake situation is much more common than one would think. One of the authors of the present work has participated in numerous sessions purporting to demonstrate all sorts of parapsychological phenomena. We can attest to the fact that it is very difficult simply to tell the truth once a false setup has been established, because the parapsychological phenomenon is a psychological crutch that is much too important for the person involved.

Crucial Choices for the Planet Are at Stake

Immensely important decisions must be made by our societies in order to face up to the inevitable consequences of the human presence on our planet. We must make choices guided, insofar as possible, by rational thought. Rationality, too, can lead to error but a lot less often than ignorance and superstition will.

Nuclear energy is an example. In chapter 3 we raised the real issue pertaining to this energy source. It's the management of radioactive wastes from spent fuels, and this is an issue that merits a debate free from propagandists exploiting fear and ignorance. The public debate should be free from the demonstrations of a violent minority for whom (in France, at least) liberty means freedom to use their Molotov cocktails to raise the costs of transporting radioactive materials to a prohibitive degree. The materials are less dangerous than are the doings of some of the activists.

Another problem, as crucial as that of nuclear energy, is the problem of hunger, which may lie in wait for billions of those human beings who will be born in the next century. This problem can be solved only by progress in optimizing the use of agricultural resources. And it must be solved; hunger makes a bad adviser.

In the past, decisions were made based on advances in agricultural techniques that often induced profound changes in the involved areas. It is obvious that this progress cannot and should not be stopped and that there is a real agro-war going on between people and nature, among countries, and among the big agribusinesses.

The United States, Canada, Argentina, China, and India have all recently authorized the growing of genetically engineered crops over a total surface of nine million hectares, including cotton, rice, wheat, potatoes, chickpeas, and cabbage. In France we've seen the actions of activist "commandos" who consider it their right to tear up plants that represent ten years of research at university laboratories. Whether they like it or not, in the end they are working *for* the supremacy of the United States. The genuine fear of the consequences of the evolution (or rather, revolution) that is under way explains their willingness to follow pseudoscientific leaders en masse, while the latter are interested only in the political power they can gain.

Openness in science is necessary but has no effect on people who have been trained in large numbers to believe in nonsense, by the joint efforts of pseudoscience leaders, the media, and lobbies. In the prologue we mentioned the genetic capital of the cavemen, the hundreds of thousands of years that separate us from them, and the gradual development of capabilities, notably those of a scientific character. We have also sung the praises of free will, that treasure that permitted the caveman to decide upon his relationships with others and with the world. But since this book is about not falling prey to self-deception, we must add that that freedom is limited by social conditions. One of the authors (G.C.) tested this by an experiment at his own expense.

During a dinner with friends, one of the young people asked the host to submit to a test: He would slowly present to the host a half-dozen calculations. The host was to concentrate on them and do them in his head. The calculations were elementary, like, "7 + 6 = how much?" These questions would be followed, without a break, by a request to name a tool and a color. The host complied with this request; amazingly, the results had been predicted in advance by a computer. Fifteen minutes later, I (G.C.) separately submitted to the same series of questions and picked the same tool and the same color. An unexpected visitor did the same. I smelled a trick and was furious at my inability to figure it out. I called Henri Broch on the spot, and Henri interrupted me to say that he knew my responses; he told me what they were, and he was right on the mark. The overwhelming majority of French people give those same responses, so they were very predictable. Where was my free will? What caused my conditioning?

Clearly, a professional with broad knowledge of a huge range of unconscious responses could gain formidable power and brush off Don Quixote's attempts at pure reason. The power of people's unconscious responses can be used for good, as by a doctor, a psychiatrist, or an artist; it can also be used in the service of detestable causes, by charlatans or demagogues. After all, this is nothing but the application of science, in a good way or bad, and science is not limited to the study of inert matter but also includes the study of people and their society.

The invocation, "May God grant you the best of health and better sense," was made by William III, king of England. He made this remark as he fulfilled one of his royal obligations, the laying on of hands to heal subjects with scrofula (a form of tuberculosis affecting the lymph glands). The remark reveals the king's irritation at having to yield to the practice of a primitive form of magic in which his people believed. His saying is still relevant for those who would like to rely on good sense to try to distinguish between good and bad alternatives. Let it not be said that Prometheus struggled in vain; science must not be snuffed out.

"Alone at night in a vast, dark forest, I had only a little candle to light my way. Unexpectedly, I came across a stranger, who said, 'Blow out your candle, you'll see much better.' The little candle is reason. It is a modest tool, no doubt, and cannot by itself solve all our problems, but at the same time it is the most valuable thing we have." The parable is Voltaire's, recounted in *The Flicker of the Candle,* by Normand Baillargeon (published in Montreal by Fides in 2001). We are tempted to qualify this affirmation of Voltaire's, for he sins by omitting all that is precious and yet is not pure reason. Don't flee from the things that attract you and enchant you. If addictive allopathic medicine, occasional homeopathy, Zen, hot springs, transcendental meditation, Taoism, whatever, is essential to your equilibrium, listen to no one but yourself; why let yourself be fooled by the clever? In this abbreviated list, there are contradictory elements, so you have to choose. But remember, only idiots never change their minds. And reason can be an excellent adviser but must be exercised the way you exercise your muscles before a long hike. Learn to avoid being fooled! This book's goal is to serve as your exercise coach in a world whose control we must not leave to shortsighted charlatans who take advantage of our honesty and ignorance.

Appendix: How to Calculate Probabilities

If you're the kind of reader who runs away whenever you see an equation, we'd advise you to skip this appendix and stop after reading chapter 5. However, if you've got some curiosity, we'd recommend that you think about the basics demonstrated here. It will be very useful to you in your life.

Coin Tosses and Probabilities

Suppose that an event has a constant probability p of occurring in a single coin toss. What is the probability of getting k of that outcome in N tosses?

If the event happens twice, for example, then the probability of exactly two such successes is $p \times p$, or p^2. If the event happens three times, the probability of three successes is $p \times p \times p$, or p^3. Thus, if the event happens k times, the probability of getting k observed successes is $p \times p \times p \times p \ldots$ (k times), so it is p^k.

This also implies that, in a series of N tosses, we must have failed the rest of the time, that is, we failed $(N - k)$ times. The probability of failing, denoted as q here, is simply $1 - p$. Remember, there are only two possibilities, success and failure, so of course the sum of their probabilities has to be equal to one, that is, $p + q = 1$, and that's why we say that $q = 1 - p$.

The probability of $(N - k)$ failures, then, must be $q \times q \times q \times q \ldots$ ($[N - k]$ times), which is $q^{(N - k)}$.

The successes can occur anywhere in the series of N trials, and this must be taken into account. So we need to consider the number of possible ways the k successes can occur among the N possible positions. This number of ways is called C_N^k and is calculated as follows: $C_N^k = N!/(k![N-k!])$. The exclamation point following a number is read "factorial" and means that you have to take the product of all the numbers from one to the number indicated. For example, 4! represents the product $1 \times 2 \times 3 \times 4$, and more generally $N!$ represents $1 \times 2 \times 3 \times 4 \times 5 \times \ldots \times (N-2) \times (N-1) \times N$. (By convention, $0! = 1$.)

In short, the probability that an event with a constant probability p will happen k times in a series of N trials is given by: $P(k) = C_N^k p^k q^{(N-k)}$. This is known as the binomial law.

In the present case, we are concerned with tossing a coin ten times, with p, the probability of getting heads on any particular toss, being $\frac{1}{2}$. The probability of getting tails is denoted by q, and it is also $\frac{1}{2}$. What is the chance of eight, nine, or ten tosses coming up the same? Here's how to do it:

Probability of getting 10 heads:
$P(10h) = C_{10}^{10}(\frac{1}{2})^{10}(\frac{1}{2})^{0} = \frac{1}{2}^{10} = 1/1,024$.

Probability of getting 10 tails:
$P(10t) = C_{10}^{10}(\frac{1}{2})^{0}(\frac{1}{2})^{10} = \frac{1}{2}^{10} = 1/1,024$.

Probability of getting 9 heads (and thus 1 tail):
$P(9h) = C_{10}^{9}(\frac{1}{2})^{9}(\frac{1}{2})^{1} = 10 \times \frac{1}{2}^{10} = 10/1,024$.

Probability of getting 9 tails (and thus 1 head):
$P(9t) = C_{10}^{9}(\frac{1}{2})^{1}(\frac{1}{2})^{9} = 10 \times \frac{1}{2}^{10} = 10/1,024$.

Probability of getting 8 heads (and thus 2 tails):
$P(8h) = C_{10}^{8}(\frac{1}{2})^{8}(\frac{1}{2})^{2} = 45 \times \frac{1}{2}^{10} = 45/1,024$.

Probability of getting 8 tails (and thus 2 heads):
$P(8t) = C_{10}^{8}(\frac{1}{2})^{2}(\frac{1}{2})^{8} = 45 \times \frac{1}{2}^{10} = 45/1,024$.

The desired probability—the chance of getting eight, nine, or ten, all on the same side—is therefore the sum of all the probabilities above, as the side required is not specified. It is appropriate to include 8, 9, or 10 tails as well as 8, 9, or 10 heads. Adding them up, we get 112 chances out of 1,024, which is 0.109375, or about 11%.

This 11% figure clearly explains the result we mentioned in the text because it means that (as surprising as it may seem) there is more than 1 chance in 10 that, if a coin is tossed ten times, at least eight tosses will show the same side.

Here's another way to put it: If a thousand people each toss a coin ten times, there will be about a hundred of them—a hundred!—who will get at least eight tosses that are all on the same side.

Probability and the Composition of a Group

There are forty members participating in the work of the organization known as STURP, the Shroud of Turin Research Project. Of these, there are thirty-nine believers and one agnostic. What is the probability that a group made up of forty people chosen at random from the ranks of thousands of American scientists would have this composition?

The binomial law, discussed in the previous section, "Coin Tosses and Probabilities," enables us to make this calculation, for which we need to know only one additional piece of information, the probability p that an American scientist is a believer. Paul Kurtz is the chairman of the Committee for the Scientific Investigation of Claims of the Paranormal (CSICOP). At an international colloquium at the University of Maastricht in 1999, he reported that a large survey showed that 60% of U.S. scientists do not believe in God, and 40% do. (The survey also showed, by the way, that the rate of belief is much less when considering only scientists at a "higher academic level.") Anyway, the probability p for our purposes is 0.4.

The probability of a group with a composition like STURP is therefore given by:

$$P(39) = C_{40}^{39}(0.4)^{39}(0.6)^{1} = 7.3 \times 10^{-15},$$

which is seven chances in a million billions.

Just for your information, here are results for some other values of p: With p of 0.25, $P(39) = 1.0 \times 10^{-22}$. No comment needed.

With p of 0.50, $P(39) = 3.6 \times 10^{-11}$. This is less than four chances out of a hundred billion.

With p of 0.75, $P(39) = 1.3 \times 10^{-4}$. Even with such an elevated probability p, the probability of such a group composition occurring by chance would be one chance in ten thousand!

Georges Charpak was born in 1924 in Poland and emigrated to Paris with his family when he was seven years old. During World War II, Charpak served in the resistance, but in 1944 he was deported to the Nazi concentration camp at Dachau, where he remained until the camp was liberated. He received a Ph.D. in 1955 from the Collège de France, Paris. Four years later, he joined the staff of the European Laboratory for Particle Physics at CERN in Geneva. In 1992, Charpak won the Nobel Prize in physics for his numerous contributions to the instrumentation used in experiments with high-energy accelerators.

Henri Broch teaches physics and zetetics at the University of Nice–Sophia Antipolis in France, where, before the creation of the Zetetics Laboratory (devoted to the scientific investigation of paranomal phenomena), he conducted research on biomolecules. He is the author of more than 120 publications and 5 books and has appeared in numerous TV and radio broadcasts. The Committee for the Scientific Investigation of Claims of the Paranormal awarded Broch its Distinguished Skeptic Award "in recognition of his innovative and creative use of new technologies in the defense of science and critical thinking."

Bart Holland, the translator, is also a researcher specializing in probability, statistics, and experimental design. He is a professor of biostatistics and epidemiology at the New Jersey Medical School.